Your Sheep

A Kid's Guide to Raising and Showing

PAULA SIMMONS
AND
DARRELL L. SALSBURY, DVM

A Garden Way Publishing Book

Storey Communications, Inc.
Schoolhouse Road
Pownal, Vermont 05261

Cover and text design by Carol Jessop
Cover photograph courtesy of *sheep! magazine,* Helenville, WI 53137
Production by Carol Jessop
Edited by Gwen W. Steege and Lorin A. Driggs
Illustrated by Carol Jessop, except drawings on page 53, which are by Elayne Sears
Technical review by William K. Kruesi
Indexed by Kathleen D. Bagioni
The name Garden Way Publishing is licensed to Storey
Communications, Inc. by Garden Way, Inc.

Printed in the United States by Capital City Press
First Printing, August 1992

Front cover: Abbie Ferris with her Suffolk-Rambouillet.

Library of Congress Cataloging-in-Publication Data

Simmons, Paula.
    Your sheep : a kid's guide to raising and showing / by Paula Simmons and
Darrell L. Salsbury.
        p.   cm.
    "A Garden Way Publishing book."
    Includes bibliographical references and indexes.
    ISBN 0-88266-770-X (hc) — ISBN 0-88266-769-6 (pbk.)
    1. Sheep—Juvenile literature. 2. Sheep—Showing—Juvenile literature. [1.
Sheep.] I. Salsbury, Darrell L., 1936- . II. Title.
SF375.2.S56   1992
636.3—dc20
                                      91-57947
                                          CIP
                                          AC

Contents

A Few Words to Parents

Any animal project that your child takes on is likely to entail a certain amount of your own time, effort, and expense. A livestock project can do much, however, to teach your child to take responsibility, to make decisions, and to be dependable and observant. These are the very skills and attitudes needed for a smooth transition from childhood to adolescence to young adulthood. The work of properly caring for sheep builds character. Perhaps for the first time, a young shepherd must accept the responsibility for the welfare of another living creature that cannot fend for itself. This responsibility develops self-confidence borne of pride in the knowledge of a job well done. Children who undertake the care of an animal attain qualities of patience and maturity beyond their peer group.

Your son or daughter will at times need adult advice and supervision, and in some instances, help with the actual work. Many jobs require two pairs of hands, and even adult shepherds need assistance from time to time.

Such groups as local 4-H clubs, Future Farmers of America (FFA), Boy Scouts, and Girl Scouts can provide much guidance. Kids can learn a great deal as they become actively involved in the many activities of these organizations. A physical disability need not prevent participation. In Georgia, for instance, a specific 4-H lamb project has been established for physically or mentally disabled youths. Each special youth has a fellow club member as a partner to assist with the care, handling, and shearing of lambs.

If you are not an experienced shepherd, your role will be made easier if a 4-H leader is supervising the choice, purchase, and care of the sheep. Other sources of advice are Agricultural Cooperative Extension agents, local sheep producers, or your local veterinarian. Where this is not possible, your child's sheep-raising project could be a learning process for the whole family. Whatever the case, enter into it in the spirit of fun and adventure. It helps to keep your sense of humor.

Introduction

One of the great things about sheep is that they are much easier to handle than a lot of other farm animals, such as cows, horses, and pigs. They don't take much room, they're fairly easy to care for, and they can be trained to follow, come when called, and stand quietly at a show. All of these features make sheep excellent animals for young people to own and raise.

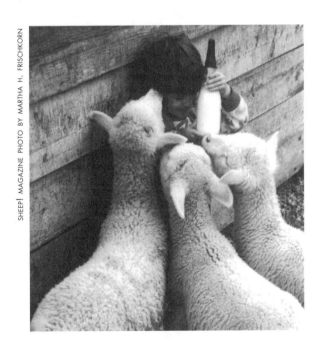

SHEEP! MAGAZINE PHOTO BY MARTHA H. FRISCHKORN

Sheep are clean, gentle, fun, easy to care for, and, above all, affectionate.

Sheep-ish History

Did you know that sheep have been the most important domestic animal in the world for over 10,000 years?

Almost as far back as the cave dwellers, people considered sheep to be one of the most important animals in the world because they could produce three things people needed for survival: milk, meat, and clothing.

For many centuries, all over the world, the traditional shepherds have been young people. The biblical King David was a shepherd as a child. Storybook characters like the "boy who cried wolf," Little Bo Peep, and Mary, who "had a little lamb," are well known to us all. Grade-school-age kids often have full responsibility for a flock of sheep. Whether it's herding them from pasture to pasture, helping with the lambing, bringing them in for the shearing, or guarding them from predators, children in many different countries have long been the shepherds of the earth.

Sheep Are Earth-Friendly

In many special ways, sheep do good things for people and for the place in which they live. And they need surprisingly little in return. Land that is too dry or too poor for growing vegetables, fruits, or grains is fine for sheep. They can eat weeds, grasses, brush, and other plants that grow on poor land.

Some people argue that it is wrong to feed grain to sheep when so many human beings don't have enough to eat. But parts of food plants like corn, rice, and wheat are too tough to be eaten by people. Sheep are able to eat those tough parts because their digestive systems are designed to handle just that sort of thing.

Sheep provide many things that we can eat or wear. Many of the world's most popular cheeses are made from sheep milk. Sheep wool, which can be used for rugs and blankets, as well as for clothing, is a renewable resource. This means that each year, right after the sheep is sheared, it begins right away to grow new fleece.

Sheep also help the soil. If you spread their manure over your garden, it will fertilize the soil, and your plants will grow strong and healthy.

If you decide to grow sheep for meat, you can also use the pelt (skin) to make clothing, and you can make candles and soap from the fat.

Getting to Know Sheep

Sheep depend on their owners for food, protection from predators, and regular shearing, but they require less special equipment and housing than any other livestock. One or two lambs or *ewes* (female sheep) can be raised in a backyard with simple fencing and a small shelter. Because sheep have thick wool coats, snow and cold temperatures don't bother them.

If you have never owned a sheep before, you probably have a lot of questions about them and about how to get your own sheep. Here are some things you might be wondering about.

Is it safe for children to be around sheep?

Sheep are one of the safest four-legged farm animals for children to handle. Most sheep are small and docile. *Rams* or *bucks* (male sheep) can be aggressive at times, but sheep, especially those that are around people every day, are usually very gentle and even-tempered. A sheep can become your pet very easily, especially if you raise an orphan lamb on a bottle.

Are sheep hard to feed?

No. Sheep don't need a lot of fancy food. They can live on grass in the summer, and hay, plus small amounts

Ewe: A mature female sheep

Ram: A mature male sheep

Lamb: A sheep less than one year old

of grain, in the winter. Fresh water, salt, and a mineral/vitamin supplement complete their diet.

Are sheep dumb?

Sheep are anything but stupid! They learn very quickly and are among the smartest of all farm animals. It's common for sheep to be able to recognize and respond to their individual names. Sheep have been mistakenly called "dumb" because of the way they naturally act to avoid danger. Sheep have no way to defend themselves. If an enemy threatens them, they cannot kick like horses, butt like cattle, or bite like pigs. Sheep can only bunch together and attempt to run away, like a school of fish, when they sense danger. Sometimes, when they are very frightened, they may be in such a hurry to escape that they run headlong into obstacles, which makes them *seem* stupid.

Do sheep stink?

Definitely not! All farm animals have their very own distinctive odor. Horses smell like horses, cows like cows, and hogs like hogs. The natural odor of sheep and their manure is not as strong as that of cattle or horses, and a lot less strong than that of pigs.

How many sheep should I get?

More than one, if possible. Sheep have a built-in social nature, known as a flocking instinct. Two sheep, therefore, would be happier than a single one away from its natural companions. The exception could be an *orphan* or *bummer lamb*. An orphan lamb is one that has been rejected by its mother or lost its mother and must be raised on a bottle. It is often just as happy around humans as it is with other sheep. Orphan lambs quickly become attached to the person who feeds them.

How old should my sheep be?

Most people start with weaned lambs (lambs that no longer need their mothers' milk). These are about two to three months of age. However, nothing can be quite as much fun (or work!) as beginning with an orphan baby lamb that you can raise from the very start. If you aren't planning to show your sheep, you may want to buy an older ewe that has been bred to lamb in the spring. These may often be purchased at a low cost, and you'll be getting two sheep for the price of one!

How much will my sheep cost?

If you are buying an adult ewe, you will find a wide range of prices. Shop around at several farms in order to learn the average market prices in your area. A nice *crossbreed* (a sheep whose parents are different breeds) will be much less expensive than a purebred or registered animal.

Lamb prices vary for the same reasons. An orphan lamb often costs very little. Sometimes the farmer with a large flock does not have the time to raise an orphan lamb and is happy to find it a good home. A weaned lamb usually costs more than an orphan.

If you don't have much money to buy a lamb, you might be able to make a deal with the owner. Offer to work for a certain number of days or hours to pay part of the price of your lamb. Almost all farms have chores that you could do. Not only will you be earning your lamb, you will have the opportunity to learn more about sheep by actually working with an experienced shepherd to help care for the flock.

The Best Breed For You

There are many breeds of sheep to choose from. A *breed* is a group of animals that have many things in common. Learning to tell sheep breeds apart is the same as recognizing the differences between breeds of dog, such as cocker spaniel and German shepherd. For instance, one sheep breed may always have white faces and long, silky wool. You may be able to recognize another breed by its black faces and coarse, wiry wool. The lambs of some breeds grow faster than those of other breeds.

Matching the Breed to Your Needs

Before you decide what kind of breed to choose, you should first think about why you want to raise a sheep and what you will be doing with it. One very good reason to raise a sheep is simply to have it as a pet. Owning a sheep as a pet is pure pleasure. More often than not a *wether* (a castrated male) makes the best pet. If it was a *bummer* (orphan lamb), better yet. Pet lambs are usually the easiest to obtain. You can choose them simply on the basis of their personality and

Breed. A type of animal that has specific inherited characteristics.

Sheep-ish History

Did you know that the "American Wild West" was first settled by shepherds and sheep, not cowboys and cattle?

There were sheep in Texas and all the western states before there were cattle.

appearance. Flock owners are often eager to sell a black lamb, one whose color is unusual for its breed, or a lamb from a set of twins or triplets whose mother may not have enough milk for all of her lambs. These can often be bought at a bargain price.

If you are interested in raising sheep for reasons other than as pets, you will want to know which breeds have the characteristics that are most important for your purposes. For instance, if you think it would be fun to raise sheep for the wool to make your own sweater, or if you want to have a way of making money by selling wool, choose a breed known for its high-quality fleece.

Breeding Ewe

If you plan to raise a ewe for breeding, consider all of the breed's characteristics. For instance, some are easy to handle, with quiet and easygoing dispositions, while others are more high spirited. Some tend to give birth easily, make superior mothers, and give birth to twins or triplets frequently. *Twinning* (the tendency to give birth to twins) is inherited. If you are buying a pet, this is not important. If you are planning to breed your sheep, however, it would be better to buy a twin, which is more likely to produce more twins. You can then have more lambs to expand your flock or to sell.

Sheep raised to produce new breeding stock should be purebred or registered animals from high-quality stock. Try to find out what breeds will sell best in your area. Do other shepherds want sheep for meat? for wool? for more than one purpose? for their unusual (*exotic*) characteristics?

Consider, too, what is already available in your area. You can then do either of the following:

- **Select a common local breed** and try to develop superior animals by giving them special care and nutrition. If your sheep win many prizes at shows, you will soon have plenty of customers for your lambs.

or

■ ***Obtain an unusual breed*** that is scarce in your area. The breed's desirable traits and appearance will attract buyers. Many producers seek out these unusual animals to crossbreed with their own flocks.

Club Market Lamb

The term *club lamb* refers to a lamb that is raised for the purpose of being shown. Club lambs are bred to have qualities that are most highly prized by judges. *Market lambs* (or *locker lambs*) are lambs raised for meat. A club market lamb, therefore, is a category of show lambs that are raised to show and then to slaughter. They should be fast growing and well muscled, in order to attain a good size by slaughter time.

When you pick out a lamb to raise for meat, you should first learn what is good *conformation* for a market lamb. Conformation is the shape and proportions of an animal. The illustration on page 8 shows some of the things that you should look for when you decide whether the sheep has good conformation. You will probably have to pay a slightly higher price for these exceptional lambs than you would for an average lamb. You will need the help of an experienced 4-H leader or sheep producer to help you with this selection.

If you raise a lamb for meat, you know that this is your purpose from the time you first get the animal, and therefore should not think of the animal as a pet. You expect to get a return from the project of raising this animal, and that return brings with it the responsibility always to treat your lamb with respect and to give

Club market lamb. *A lamb raised to show and then to slaughter for meat.*

Wether. *A castrated male sheep.*

Bummer. *An orphan lamb that is fed from a bottle rather than nursed from a ewe.*

it the best care you possibly can, right up to the time it must be slaughtered.

Tips on Buying a Club Lamb

- Buy a healthy lamb. Ask your veterinarian to examine the animal for signs of disease.

- Ask for and use good, unbiased advice. (Remember that everyone has an opinion, but not everyone is correct.)

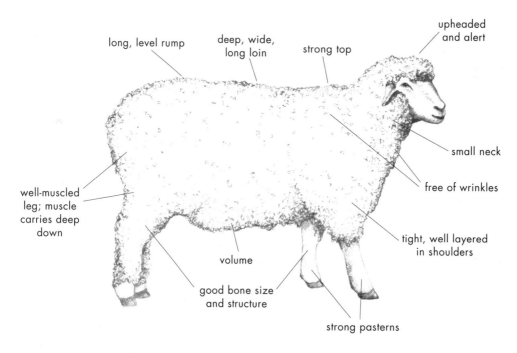

This sheep has good conformation.

- Price is not always a reflection of quality. The most expensive lamb is not necessarily the best lamb. Learn to recognize the finer points of conformation (see drawing) and what traits the judges are looking for.

- Championships are not bought, they are earned. The most expensive lamb with the best potential show traits will not be a winner without hours of training, *fitting* (washing and clipping a lamb for showing), and your ability to show it to best advantage.

- Most faults don't get better with time. If the lamb has faulty conformation, its defects will become more pronounced with age.

- Don't buy what you cannot see. If you are considering an older lamb with long fleece, you may not be able to see its body form. Learn how to determine desirable traits, such as a thick leg muscle, by feeling the lamb.

- Spend only 80 percent of what you think you can afford. You will always have some expenses you hadn't planned for.

- If you need a show lamb, an inexpensive lamb with poor conformation is not a bargain. Keep shopping until you can find quality at a price you can afford.

- Look before you leap. Don't be rushed into making a purchase until you feel sure it is the right decision.

- Sheep are happier with companions. Try to buy two lambs to raise together. By the time you are ready to show them, you can choose the one more likely to win a prize.

Sheep Breeds and Climate

Where you live will also determine which is the best breed for you. You should choose a breed that can

knock-kneed and
splay-footed

bent leg

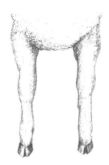

correct

Poor conformation includes knock knees and bent legs.

survive in cold weather if you live in an area with severe winters. If you live in a very wet area, look for a breed that tolerates rainy weather. If you live in a desert-like area, you will want one of the breeds that is adapted to hot, dry climates. Look around and see what breeds of sheep are being raised locally — these may also be the best breeds for you.

Exotic, Rare, and Minor Breeds

Sheep breeds that are rare (small in numbers) or minor need to be saved from extinction. Many of these rare breeds were important at one time but have been replaced by more productive types. This doesn't mean that the old-fashioned breeds should die out. They may have important characteristics, such as the ability to milk well in hot weather or resistance to worms. There is interest today in preserving all breeds of sheep and other livestock, even though we are not sure how they can be used in a large farming operation.

Exotic breed. A breed of sheep that is strikingly unusual in appearance or temperament.

Exotic breeds, just as their name says, are strikingly unusual in appearance and/or temperament. Some have strange colorings and markings. Others have "hairy" wool. Still others have four large horns. Some are quiet and docile, and others can be quite aggressive and hard to handle and/or confine. Some are expensive and hard to locate. You can get a great deal of pleasure from some of these exotics, however, and they can also be profitable. But before you choose an exotic breed, take extra care to investigate the sheep's temperament and appearance.

Knowing the Terms

■ **White or black face.** These terms mean exactly what they say: they describe the color of the wool on the sheep's head and face. Normally, the wool on the lower legs is the same color as that on the face.

- ■ **Open or closed face.** These terms are used to describe how much long wool is on the sheep's face. An open-faced sheep has only very short, hairlike wool on its face. A closed-faced sheep has long wool on its face. On a closed-faced sheep, the wool may grow all the way down to the animal's nose. Too much wool around the eyes causes the sheep to be "wool blind." The excess wool must be clipped away so the animal can see.

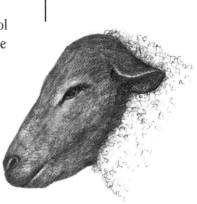

open and black face

- ■ **Prick or lop ear.** Just as a German shepherd's ears stand up and those of a cocker spaniel hang down, sheep's ears can stand straight up (prick ear) or hang floppily down (lop ear); some even stand out to the side.

- ■ **Polled or horned.** A polled sheep has no horns. A horned sheep has horns.

closed and white face

You can recognize sheep breeds by the traits they have in common.

Some Popular Breeds

There are so many breeds of sheep that it would take a whole book to describe them all. Here are some of the most popular breeds, as well as a few minor breeds. If you keep in mind your reasons for owning sheep, these brief descriptions will help you know which breeds are right for you.

Dorset

Open face (usually)
White face and legs
Both polled and horned types

This breed is considered one of the best choices for a first sheep. The Dorset is medium-sized, with a very gentle disposition. The first Dorsets were bred in England. A Dorset has very little wool on its face, legs, and belly. This trait is an advantage you will learn about in the lambing chapter.

Dorsets are a fine choice, especially if you are interested in both wool and meat. Their lightweight fleece is excellent for handspinning. They have large, muscular bodies and gain weight fast, making this breed an excellent choice for market lambs. They are also extremely good mothers, so would be a good choice if you want to breed sheep. They are one of the few breeds that can lamb in late summer or fall.

Hampshire

Partially closed face
Black face and legs
Polled

Like the Dorset, the Hampshire was first bred in England. It is among the largest of the meat types, and the lambs grow fast. The wool on a Hampshire sheep extends about halfway down its face. They have very gentle temperaments, and are often selected for club projects for this reason.

Suffolk

Open face
Black face
Polled

Suffolks are very similar to Hampshires. They, too, are large, with fast-growing lambs. They have less wool on their faces and legs than do Hampshires. Because Suffolk lambs grow fast and are very popular, they are an excellent choice for club projects. Like Hampshires, they are normally gentle, but some Suffolks can be somewhat headstrong and difficult for younger children to manage and show.

Emily Jackson with her Hampshire ewe, Shirley.

Amanda Alford with her Suffolk sheep.

Columbia, Corriedale, Polypay, and Romney

All of these breeds are good wool producers. Because they have very gentle temperaments, they make excellent friends and pets. Sometimes individual sheep within these breeds can be quite large. They are white-faced and polled; some are open- and some are closed-faced.

The Columbia breed was developed in the United States in 1912. They are large sheep but are calm and easy to handle. They produce heavy market lambs and a heavy, dense fleece.

Dorsets and Corriedales at Wooly Hill Farm (Lincoln, VT)

Polypay twins and their mother

The Polypay was developed in the United States as a dual-purpose (meat and wool) sheep. Polypays have a high twinning rate, good mothering ability, medium-fine wool, and a long breeding season.

The Romney originated in the low, wet lands near Kent, England, and so they are suited to areas where it is cool and wet. They were imported into

Romneys are quiet, calm, and gentle.

the United States in 1904. They have black *points* (noses and hooves) and a long *staple* (length of wool), soft fleece that is ideal for handspinning. The average fleece weight is 10 pounds. They also produce good market lambs. Romney ewes are quiet, calm mothers.

Tunis

Closed face
Reddish-tan face and legs
Lop-eared
Polled

We know that the Tunis breed has been around for over 3,000 years, making it one of the oldest sheep breeds. It is called a minor breed because there are relatively few of them in the United States. They are medium size, hardy, docile, and very good mothers. The reddish tan hair that covers their legs and faces is an unusual color for sheep. They have long, broad, free-swinging,

Sheep-ish History

The first Tunis sheep in North America were imported from North Africa in 1799. A flock was maintained at George Washington's home at Mount Vernon. The breed was popular in the South until the Civil War, when many sheep were destroyed.

Jim Lillie's Tunis sheep, in Fitchburg, MA

lop ears. Their medium-heavy fleece is popular for handspinning. The Tunis thrives in a warm climate, and the rams are able to breed in very hot weather. The ewes often have twins, have a good supply of milk, and breed over a long period of time.

Katahdin

The Katahdin breed of sheep is an easy-to-care-for, meat-type sheep that has hair instead of wool. It does not require shearing because it sheds its hair coat once a year. It is able to tolerate extremes of weather. Except for the fact that Katahdins do not produce wool, they possess all of the ideal traits for the pet or small flock owner: They are gentle with mild temperaments, require no shearing, have few lambing problems, are excellent mothers, and have a natural resistance to parasites.

KATAHDIN HAIR SHEEP INTERNATIONAL

Katahdin sheep have hair instead of wool.

Buying Your Sheep and Bringing It Home

CHAPTER 3

Once you have decided on what breed of sheep you want, you can choose the individual animal you want to buy. You may need some help with this decision. It's often a good idea to have the animal you've chosen examined by a knowledgeable person, such as an adult sheep owner or a veterinarian.

It is best to buy your sheep directly from the person who raised it. You can ask questions about the sheep's history and see the flock that it is coming from. Avoid buying a sheep at an auction, because you don't usually have the opportunity to talk to the owner, and you may also be rushed into making your decision.

Let's talk first about what would make an ideal sheep. Above all else, your sheep should be healthy. You can get a good idea about this just by looking at the sheep carefully and observing the following:

- It should seem alert and thriving.

- It should come from a flock with no major medical problems. If the whole flock is in a good state of health, chances are good that your sheep has been

Qualities of a Good Sheep

- Good conformation, free of obvious defects

- Rear legs plump and well muscled

- Teeth meeting dental pad well; jaw not overshot or under-shot

- Strong *pasterns* (ankle joint just above the hoof)

- Rams: have fertility checked by a veteri-narian

- Ewes: good udder with no lumps or damage

well cared for, has a good family background, and has not been exposed to diseases or parasites.

- Learn to recognize the ideal conformation of a sheep (see illustration, page 8). While you may not find or be able to afford a "perfect" lamb, see how close the lambs are to the ideal conformation. The sheep should be the normal size and weight for an animal of that age and breed. If it has large bones, it will have more meat, and if a ewe, will handle pregnancy and birth more easily.

Judging a Sheep's Physical Condition

Examine the sheep's eyes, teeth, feet, and other body parts, including its fleece. Look for signs of good health and conformation, and avoid taking a sheep with any of the problems described below:

Eyes. Runny or red eyes or damaged eyes may mean the sheep is diseased.

Teeth. Worn or missing teeth will interfere with eating.

Jaw. The lower jaw should not be undershot or overshot (see illustration on page 20).

Head and neck. Make sure there are no lumps or swelling under the chin. These may be related to a disease known as *bottle jaw,* which is caused by an untreated worm infestation.

Feet. If the sheep is limping, it may be injured or it may have footrot (see pages 46-48). Even if the sheep seems healthy, notice whether other sheep in the flock are limping. This may mean the flock has footrot or other foot problems, and your sheep, too, may be infected. Untrimmed or overgrown hooves show that the hooves haven't been cared for properly.

Body conformation. Sheep with narrow, shallow bodies tend to be light muscled. Sheep with wide backs and deep bodies are more desirable.

Body condition. Don't choose a ewe that is extremely thin. Exception: A ewe that has just raised lambs may be thin, and this does not mean she has a health problem. Don't choose a very fat ewe, either. She may have trouble *lambing* (giving birth).

Potbelly. A thin lamb with a "potbelly" usually has a heavy infestation of worms.

Udder. If you are buying an adult ewe, check the *udder* (organ containing the milk glands and nipples). If it is lumpy, she may have had *mastitis* (an udder infection), which sometimes causes her to have no milk for future lambs.

Tail. An extremely short tail may mean that the *docking* (tail shortening — see pages 39-40) was not done properly. This is a serious defect, causing the whole area to be weakened and possibly to create problems during lambing.

Manure. Runny droppings or a messy rear end could mean that the sheep is sick or has worms. Exception: Sometimes a lamb that has been grazing on new, lush spring grass will have runny droppings, and this is not a health problem.

Wool covering. Each breed has a certain amount of wool on its face or legs. Your sheep should have the correct amount for its particular breed. Avoid sheep with excessive wool around the eyes.

Fleece. A ragged, unattractive fleece could be caused by disease, keds, or lice.

Conformation. *The shape and proportions of an animal.*

Poor: Undershot jaw

Poor: Overrshot jaw

Good

Check for good jaw conformation.

External parasites. Look for any other signs of external parasites such as keds or lice.

Judging the Age of a Sheep

The front teeth of sheep are very similar to the front teeth of humans, with one big difference: sheep have front teeth only on the bottom. Where you would expect to find upper teeth, a sheep has a hard gum line, called a *dental pad*.

You can estimate the age of sheep by examining their teeth. The front teeth of sheep begin to show at two to three weeks of age, starting with one pair in the center, followed by three more pair a few weeks later. The lamb will have eight teeth in all. Sheep lose these baby teeth just like children lose their baby teeth. The baby teeth are then replaced by permanent teeth. When the sheep is about one year old, the first pair of permanent teeth appears. Each year after that until the sheep is four years old, it gains another pair of permanent teeth. When the sheep is four years old, all four pair of baby teeth have been replaced with permanent teeth.

What to Ask the Sheep's Current Owner

Here are some important questions that only the seller can answer. Don't be afraid to ask them.

What vaccinations or treatments has the sheep had, and when?

Young lambs may have received some, but not all, of their primary vaccinations. In order to complete the health program, you must know where to start. The sequence and timing of the vaccinations are important.

You may be purchasing a lamb in the middle of its immunization schedule. You will need to know what shots were given and when, so that you can get the rest of the shots right on schedule and with the same product. Finding out about your sheep's early care is an excellent opportunity to become more informed about vaccines. For more about vaccinations, see pages 41-43.

When was it wormed and with what?

This is another important piece of your lamb's health record. You will need this information in order to maintain the scheduled wormings with the same product that the lamb has already received. For more about worming, see pages 43-44.

What kind of feed was it normally given?

Any sudden change in the kind of feed you give your sheep can make it sick. Digestive upset can also slow down your sheep's growth or even cause its death. If you are not sure what to feed your sheep, continue giving it the same kind and amount of feed it has been given up until now. If you are buying your sheep from an experienced shepherd, take advantage of his or her experience. Ask where to buy the sheep's feed and how much it will cost, as well as what medications and other supplies you may need. For more about feeding, see pages 33-38.

Was it a twin? Was its mother a twin?

Twinning (giving birth to twins) is a highly desirable *genetic trait* (a characteristic passed along by the lamb's parents). The possibility of twinning is mostly determined by the ewe. If the ewe is a twin, she is more likely to give birth to twins of her own. Even if you are buying a ram, ask if he was a twin. If so, his daughters are also likely to give birth to twins.

**Did you know that long
ago a person could be
put in jail for selling a
sheep?**

About 5,000 years ago,
people began to spin
wool and use it for
clothing. Wool made
such excellent cloth that
during the time when
Europe was ruled by
kings, the development
of special breeds of
sheep was extremely
important to the welfare
of the kingdom. In fact,
subjects could be
thrown in the dungeon
or put to death for
selling a special breed of
sheep to someone
outside the kingdom.

If the lamb is an orphan, why does it have no mother?

A very good lamb may be pushed away by the mother
because she had triplets or quadruplets, and this was
just one too many for her to care for. But if the mother
was simply a "poor mother" (some individuals are not
as interested in mothering as others are) or had no
milk, she might have passed these traits along to her
daughter. The little female bummer will then be less
valuable for breeding when she grows up, because she,
too, may be a poor mother or not have enough milk.

Has the flock had a history of medical problems?

You don't want to buy a bunch of problems. Sheep
diseases spread quickly among the flock. A sheep from
a sick flock may have been exposed to a number of
diseases, especially foot problems. Notice whether any
of the flock are limping or on their knees when grazing. If their feet show signs of neglect, such as overgrown hooves, be very cautious about buying from this
source. Similarly, you don't want to buy sheep that
may have lice, keds, or problems with internal parasites. Health problems are discussed in chapter 6.

Is there a record of this lamb's growth?

Ask for the date of the lamb's birth, its weight at birth,
the date of its weaning, and its weight at weaning. You
can use this information to figure out the rate of
growth. Rate of growth is more important than the
size of a lamb. Two lambs of equal weight can often be
two or more months apart in age! If you are raising a
lamb for market, you shouldn't purchase a slow-
growing runt, because it won't develop enough weight
to bring a good price.

After you buy your lamb, continue to weigh it
frequently. Keep track of this information on your
sheep record. (See page 42 for sample record.)

Getting Ready to Bring Your Sheep Home: Fences, Shelter, and Food and Water

Fences

Before you get that first lamb, you must prepare a place to keep it. This means building fences if you have none, or checking existing fences to be sure that they are secure. *The purpose of a fence is to keep the sheep in and to keep dogs (and coyotes) out.* A sheep running loose can be hit by a car. A dog playing with the sheep in the pasture can kill it or cause it serious injury. The fence is the first priority.

Of the many types of fence, the best for sheep is a smooth-wire electric or a nonelectric woven-wire fence. Some nonelectric wire fences have six or more tightly stretched wires and heavy posts, but woven-wire "sheep fence" is better. This fencing will keep sheep in, but to keep predators out, put electric wire on the outside or a couple of strands of barbed wire above and one below.

The strength of the electric fence is in the shock. Well-made, electric fence chargers (*energizers*) are available at farm supply stores and fence supply dealers. Most of these chargers will operate on a few dollars of electricity per month.

Plan on practicing *pasture rotation.* This means that you control where and for how long your sheep graze. You do this by *cross-fencing,* or dividing, your large pasture into smaller *paddocks* (small pasture areas) with inner fences. The outer fence *(perimeter fence)* must be permanent and very sturdy in order to keep dogs out of the pasture, but the inner fences can be any inexpensive woven-wire or portable electric fencing, which is quick and easy to construct.

PHOTO BY RUTH RICHARDSON

Woven-wire fence does a good job keeping sheep in.

Pasture rotation is better for grass growth as well as for sheep health. The grass can grow in the empty paddocks while the sheep graze in another paddock. The sheep get smaller pastures at one time, but the grass is always fresh. For best results, make several paddocks out of your total pasture and move the sheep every few days during the grazing season (spring, summer, and fall).

Shelter

Keep young children away from an electric fence. The fence shock is painful, and a small child may get trapped between the wires.

Lambs and sheep require shade from the heat of the summer sun and shelter from the rain in summer and from the snow and wind in winter. Sheep are happiest in a three-sided shelter. Sheds or barns should be well ventilated and kept clean. If you do not have a shed or barn, portable metal shelters used by pig producers make a nice shelter for one or a few sheep.

Food and Water

Beginning the day you bring your sheep home, you will need to provide it with hay or put it in a good pasture. It will also need fresh water. While the feeding arrangement can be quite simple at first, making sure there is plenty of fresh water is really important. For a few sheep, a washtub works very nicely. Do not use a bucket as it is easily tipped over. For more on feeding, see pages 33-38.

Getting Along With Your Sheep

A sheep's most powerful instinct is, at all costs, to avoid being trapped. A cat has claws, a porcupine has quills, a skunk has its scent, but a sheep has only one defense from danger — escape!

This drive to avoid being trapped can get sheep into a lot of trouble. It can also be a major frustration for a new shepherd. A sheep's brain, like that of its parents and grandparents, is programmed in this "escape mode." It has been that way for hundreds of centuries, and it cannot be changed. This is why dogs in the pasture can cause so much trouble. When the sheep see the dog, they run to escape. When the sheep begin running, the dog thinks it's fun and begins chasing them. It's a vicious circle, since if the sheep didn't run, the dog wouldn't chase them.

To be a successful shepherd, you must learn to take advantage of the sheep's way of thinking. Sheep will come to you if you don't do anything to startle them or make them think that you are trying to chase them. For instance, if you try to drive the sheep into the barn, they will avoid going in if they think you are trying to trap them. But, sheep love to eat, and they will do almost anything to get their favorite treat. They especially love peanuts, apples, and grain. If you show them a bucket of grain and use it to coax them to follow you, you can lead them into the barn very easily. In fact, they will be so eager to follow you that they will actually get pushy.

Allow them time to eat the grain before you try to gather them tighter or try to catch one. The grain is their reward for coming in, and eating the grain should be a good experience for them. Once the sheep are indoors or in a pen, use lambing panels or a gate to squeeze them together so they can't run away. Walk up to the group, catch one sheep, and move it to a clear area where you can handle it.

If you say your sheep's name and offer it a treat at the same time, you can soon train it to come when you call its name. You will be surprised how quickly it will learn. If you sit quietly on a small stool or box in the middle of the sheep pen, your sheep will be very curious about you and begin to approach. If you remain still and talk softly, it will soon come close enough to smell your hair and nibble at your clothing.

As your new sheep becomes tame, usually it will want to be scratched and petted. *Never* attempt to scratch it on the top of its head or nose. Sheep don't like this. They prefer to be stroked under the chin and scratched on their chest between their front legs. Once in a while, you will find a sheep that likes to have its back and/or rear scratched. I have a *bellwether* (the leader of a flock) named George who likes to have his rear scratched. He will come up, look me in the eye, go "baa-a-a," turn around, and wiggle his tail. This is his way of saying, "I want my rear scratched."

Handling Sheep

The time will come when you will need to make your sheep do something that it doesn't really want to do. There are ways to make your sheep do what you want without frightening it or losing its affection and trust. Although calm, adult sheep that are used to people can be trained, it's easier to begin training as early as possible. Even new lambs should get used to being touched by people. A lamb that has not been handled very much is practically impossible to catch in a pasture! Don't begin training until the lamb begins to nibble at grass (about two weeks of age). The ewe and her lamb need time to form a strong bond (see *mothering up,* page 65) during the first couple of weeks after lambing.

Dock. *Rear portion of a sheep.*

Don't hold a sheep by its wool. You don't like someone to grab your hair, and that is exactly what you would be doing if you tried to lead, hold, or restrain your sheep by pulling its wool. It hurts! To make your sheep stand still, place one hand under its chin and the other hand on its hips or *dock* (tail area) and slightly to the rear. When held this way, it can't go forward or backward, or swivel away from you. Control your sheep with your hand

Control your sheep with one hand under its chin and the other hand on its dock.

gripped firmly under the chin, and walk it forward by giving it a gentle squeeze on the dock. As the sheep starts moving, it's a simple matter for you to keep up with it. When you want it to stop, hold it back with the hand that is under its neck.

Halters

When your lamb is about one month old, you will probably want to train it with a *halter* (a strap that fits over the sheep's nose and head). Your sheep will not like its halter at first, but with a little practice and repetition, it will soon settle down. Put the halter on as shown in the photo below. After it is in place, grip it close to the head and place your other hand over or behind the sheep's hips, just as you learned how to lead the sheep without a halter. Start leading your sheep by pulling on the halter and pushing on its rear. Have someone help you push, if you wish. Be patient. A lamb may buck and fuss a bit, but it will soon get used to working with the halter. You will find directions for making halters on page 84.

PHOTO BY LESLIE IRVIN

Practice training your sheep with a halter.

The Shepherd's Chair Hold

When you have to give a sheep shots, trim its feet, or shear its wool, you can take advantage of a reflex all sheep have: Once all four of their feet are off the ground, they can be placed on their rump and they will sit still.

To get your sheep into this sitting position, slip your left thumb into the sheep's mouth in back of the incisor teeth, and place your other hand on the sheep's right hip. Bend the sheep's head sharply over its right shoulder, and swing the sheep toward you. Lower it to the ground as you step back. From this position you can lower it flat on the ground or set it up on its rump for foot trimming.

It's important to withhold their feed for several hours before handling or shearing sheep. This sitting position is uncomfortable for sheep with a full stomach.

Health and Happiness: The Basics

You want your sheep and lambs to be happy. Just remember that many of the same things that make you happy will make them happy.

Sheep in this sitting position will hold still for you.

- **Regular feeding.** Do you like to eat at a regular time? So do sheep. Try to feed your sheep at about the same time every day. When sheep are fed grain and/or hay in addition to pasture, they are usually fed once in the morning and then again in the evening (see pages 33-38).

- **Customary diet.** Sheep cannot adapt to sudden changes in either the type or amount of feed. It could make them very sick. If you are going to

American yew

Tansy ragwort

Oak

Lupine

Rhododendron

change feed, do so gradually. Start adding the new feed and reducing the old feed over a period of about 10 days.

- **Fresh water.** Sheep like *fresh* water. Water is an important part of good health, and they will drink more of it if it is fresh. If manure gets into the water, empty the tank and clean it out before refilling.

- **Hay.** Feed your sheep a green, leafy grass or alfalfa hay in the winter or when the pasture is short or covered with snow. *Never* throw the hay on the ground, because it will become dirty and wet, and contribute to worm infection. Put it into a feeder of some type.

- **Salt, minerals, and vitamins.** A good supply of salt and a mineral/vitamin supplement is important for good health. These supplements are usually placed in sturdy wooden boxes that won't tip, or in a hanging feeder. (For instructions on how to make your own mineral feeder, see page 91.) The salt should be "loose" or "granulated." Never feed your sheep salt blocks intended for cattle. The sheep may attempt to chew the block and harm their teeth. The same is true for the mineral/ vitamin supplement. Be sure that it is the "loose" or "powdered" type made especially for sheep and *not* for cattle. *Cattle minerals often contain levels of copper that can be toxic to sheep. Never feed cow or pig supplements to sheep!*

- **No toxic plants.** Watch out for possible sources of "food poisoning." If you allow your sheep to play or graze on the lawn, make sure they do not eat any ornamental plants or shrubs. Sheep are curious and love to nibble at strange plants. Your County Extension Service can tell you what poisonous wild plants grow in your area. Some are shown here.

- **Grain storage.** Sheep *love* grain, and they will figure out how to get into the grain storage area, if possible. Keep your grain in a safe container that cannot be reached or knocked over by the sheep. If they overeat grain, it can cause severe stomach upset and possibly death.

- **No dogs allowed.** Keep your sheep safe. *Never* allow your dog or any other dog to play with the sheep. Sheep are frightened by dogs, and a dog that is barking at them through the fence or running in the pasture causes severe stress in the sheep, which can lead to over-exertion, heat stress, and heart failure.

- **Shelter.** Provide clean, well-ventilated shelter. If you have several sheep, you will need about 15 to 20 square feet per adult animal. Good ventilation (but not drafty) is mandatory. One side of the shelter open to the south provides ventilation necessary for good health.

- **Safe at night.** Even good fencing can't always provide enough protection at night. The shelter should be located within a corral or small pen constructed of tight-woven wire or "cattle panels" that will positively keep any dogs or other predators from getting to the sheep. If you place feed in the shed each evening, the sheep will naturally go inside for the night.

- **No parasites.** Make sure that your sheep are free from sheep keds (ticks) and lice. If you are not sure, have your sheep examined by your veterinarian. These skin parasites are easily treated, and once you get rid of them, your sheep can only catch them again from another infected sheep.

- **Foot care.** Hoof trimming usually needs to be done only once or twice a year. Sore, overgrown hooves are not comfortable. (For more about hoof care, see pages 46-48.)

PLANTS POISONOUS TO SHEEP

Nightshade

Sheep laurel

Milkweed

Ragweed

- ***Treatment for worms.*** Sheep and lambs pick up worms from the pasture and they should be treated at least twice a year (more often, if you have many sheep on a small pasture). Consult with your veterinarian or Extension Service for instructions as to the type and use of de-worming medications. (For more about worming, see pages 43-44.)

- ***Shearing.*** Learning to shear your own sheep can be fun, and you may be the best person to give your sheep a "haircut," because you know them well and can take time with them. Professional shearers, however, usually have good training and experience. (For more about shearing, see pages 49-54.)

- ***Shade.*** Shelters are not just for cold or rainy weather. Sheep need shade in the summer. An open-sided shed, shade trees, or a canopy roof can give the needed cooling shade.

- ***Training.*** Work with your sheep or lamb. Even if you don't plan to take your animal to livestock shows, you will both be happier if you have done some constructive training. Start working with your lamb or sheep as early as possible. Once you have earned its confidence, it can be trained to follow, come when called, and stand still. Peanuts and small bits of apple make great rewards and treats.

- ***Pasture rotation.*** Sheep love new, clean grass. You can provide this for your sheep by practicing *pasture rotation* (see pages 23-24).

Feeding Your Sheep

How Sheep Digest Their Food

Sheep belong to a large class of animals called *ruminants*. This family includes such animals as reindeer, buffalo, elk, moose, deer, cattle, and goats. Ruminant animals *regurgitate* (throw up) a wad of feed called the *cud*, and re-chew it. This is called "chewing the cud." Many people say that ruminants can digest grass and hay because they have four stomachs, but this is not correct. Most ruminants have four compartments in the upper part of their digestive systems that work together to break down plant material such as fresh grass and hay. In the three "fore-stomachs" *(rumen, reticulum,* and *omasum)* the feeds are mixed together and fermented by microorganisms, and in the one true stomach *(abomasum)* the fermented feed is finally broken down by chemicals called *enzymes*. This large digestive tract with its separate compartments helps ruminants ferment and digest a wide variety of feeds, including flowers, leaves, straw, young twigs, soft bark, apples, grain, and pine needles.

Ruminants cannot adapt quickly to big changes in their diet, such as a change from feeding on grass to feeding on hay. They depend on microorganisms in the rumen to ferment their feed. The microorganisms that help to digest the grass and hay are different from the ones that digest grain. Therefore, you must make diet

Ruminants. *Animals whose digestive systems make it possible for them to feed on grass and hay.*

Equipment Needed for Bottle Feeding

- **Bottles and nipples.** Start with an ordinary baby bottle and nipple with the hole slightly enlarged. When the lamb is a week old, use a lamb nipple on a soda bottle.

- **Lamb milk replacer.** One 50-pound bag should be enough to raise a single lamb.

changes slowly, to give the microorganisms time to grow in numbers, so they can digest the new feed. It may take a week or more before a hay-fed animal can digest a large amount of grain.

Let's look at the feeding requirements of sheep, starting with lambs, both orphan lambs that have no mother and nursing lambs.

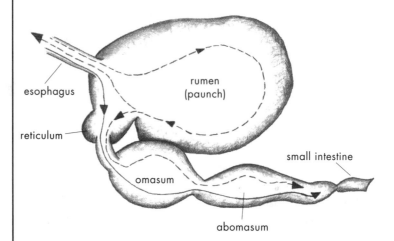

The pathways taken by a ruminant's feed

Bottle Feeding the Orphan Lamb (from Birth)

Feeding the orphan lamb is a lot like feeding a newborn human baby. (Page 22 explains why a lamb might be orphaned.) The key to trouble-free feeding is to offer a small amount of milk four to six times a day. Feeding small amounts helps prevent digestive problems such as *colic* (trapped gas) and *scours* (diarrhea).

1–3 days old. Begin by mixing *lamb milk replacer* (available at feed stores) with *twice* the water called for on the label. Feed approximately 4 ounces per feeding, four to six times a day. Hold the bottle *vertically*, with the nipple pointing almost straight down, so that the lamb has to stretch upward in order to nurse. This is the position the lamb would be in if its mother were nursing it, and it prevents the lamb from swallowing air. In the natural nursing position, the baby lamb's nose and face are somewhat buried under the ewe's hind leg and udder. Some lambs respond to bottle feeding better if you hold the bottle close to the nipple between your thumb and forefinger, so that your free fingers lie loosely over the lamb's nose. This provides a "security blanket" of warmth and closeness that feels natural to the lamb and creates a bond with you, like the bond between a ewe and her lamb.

3–7 days old. At about three days of age, begin to increase the amount of milk replacer powder in the formula, until you are mixing it full strength (as per label) by about the fifth day.

7–14 days old. By this age, the lamb will outgrow the human baby bottle. You should get a regular lamb nipple and place it on a 12-ounce beverage bottle. Continue to slowly increase the amount of milk per feeding. At a week of age (depending on the size of the lamb), a healthy lamb should be taking about 4–6 ounces of formula per feeding. As the lamb gets older and takes more milk at each feeding, reduce the number of feedings to three per day.

About 10 days old. At this time, the lamb will begin to show curiosity and interest in solid foods by nibbling at grass, hay, and grain. You should provide a small amount of commercial baby lamb feed, called *creep feed,* or finely ground, 15 percent protein supplement and water. It is at this age that the lamb's digestive system actually begins to develop in response to the roughage that it starts to eat. The commercial creep feeds usually contain small doses of medications that help to control common lamb illnesses.

Katie Grotts, feeding grain to her sheep Big Girl.

Creep feed. *A feed that is specially formulated for baby lambs.*

15–30 days old. Continue to feed approximately 6–8 ounces of milk per feeding, three times daily. During this period the lamb will begin to eat more creep feed and leafy alfalfa or good grass hay, such as brome.

A creep allows small lambs in and keeps mature sheep out.

Scours

■ Overfeeding may cause a yellow, pasty diarrhea, called *scours*. If this occurs, reduce the amount of milk you feed the lamb. Using water instead of milk for one feeding every 24 hours also helps reduce the diarrhea, and assures that the lamb gets all the water it needs. Simple aids such as a few teaspoons of Pepto-Bismol or diarrhea medication for human infants can be used for baby lambs.

■ Unclean bottles can also cause scours. Bacteria may be present in any container that is not perfectly clean. Bottle lambs are more prone to infections and stomach upsets at an early age than lambs nursing their mothers. Should the diarrhea turn a whitish color, get help from your veterinarian immediately. If this is not possible, begin antibiotic therapy immediately. You can use one of the diarrhea medications for young calves or swine that

Scours. A yellow diarrhea in lambs caused by overeating or unclean bottles.

Helpful Hint

Lambs prefer fresh food and water. Feed small amounts of fresh hay, grain, and water. Throw out what the lamb doesn't use and replace it with fresh food and water daily.

are available at most feed or animal health stores. The dosage of these medications is determined by the weight of the animal.

The Nursing Lamb

The feeding schedule for the nursing lamb is identical to that for the orphan lamb, except that you don't have to bottle-feed your animal. Observe the lamb and check the ewe, especially during the first week, to make sure the ewe is producing enough milk. Lambs that are receiving enough food will be alert and lively. When they sleep, they sleep soundly and stretch when they wake up. Weak lambs are dull and listless. Ninety percent of lamb deaths during the first week of life are from chilling and starvation, not disease! The starving lamb will *rarely* cry out after the first day of hunger, so it's important for you to observe carefully.

Begin creep feeding when the lamb is about ten days old, just as you would for an orphan lamb. Feed the same type of hay as is being fed to the ewe. The lamb will often follow her example. Mixing some of the ewe's grain mixture with the creep feed will also stimulate the lamb to begin eating solid food. You can offer a small amount of salt and mineral/vitamin supplement, too. Lambs from about two weeks old are quite curious and nibble just about anything within reach.

Weaning Orphaned Lambs

Weaning is the process of getting lambs used to eating pasture or hay and creep instead of milk for their nourishment. Depending on their size, orphan lambs can be gradually weaned from milk by starting to

A nursing lamb and its mother

Weaning. *The process of changing the lamb's way of feeding from nursing to eating other food.*

reduce the frequency of feedings beginning at six to eight weeks of age, if they are consuming adequate amounts of pasture or creep and hay. If your lamb enjoys and finishes the feed you give it, you will know that you can start weaning.

Weaning Nursing Lambs

The age at which nursing lambs are ready for weaning varies greatly from lamb to lamb — anywhere from six to sixteen weeks old, but usually about twelve to fourteen weeks. Separating the young from their mothers during weaning can be very upsetting for everyone, including the lamb, the ewe, and the shepherd! It is best to leave the lamb in its familiar pen and remove the ewe to new quarters. If you don't have enough space to separate them completely, remove the ewe during the daytime hours and reunite them during the night. This results in less nighttime crying that disturbs the shepherd's sleep. Forcibly weaning a lamb from its mother is not particularly necessary in a small farm situation. As the ewe's milk supply declines normally, she does her own weaning by not allowing the lamb to nurse as long or as often.

Adults (Ewes and Rams)

Mature or adult ewes and rams require only enough nutrients to maintain body functions. They do not need extra feed to grow, since they have reached their full size. Normally, their feed consists of grass pasture, salt, a mineral/vitamin supplement, and water. As long as there is enough forage and water, adult sheep will be properly fed. In periods of drought or when the grass becomes short, it is necessary to supplement their diet with hay or grain. For information about feeding pregnant and nursing ewes, see pages 61-62, 70, and 78.

Normal Health Care, Treatment, and Medication

Sheep, both lambs and adults, need a certain amount of grooming, health care, and medical treatment, just as people do. This chapter explains the basics for keeping your sheep in good health and for dealing with health problems when they come up.

Procedures for Lambs

Why do I need to dock (shorten) my lamb's tail?

Lambs' tails are cut short for health reasons, *not* just for appearance. Long tails are very woolly and may accumulate large amounts of manure, which attracts flies and maggots. Tails interfere with breeding, lambing, and shearing.

Docking. *Shortening the lamb's tail.*

How do we dock tails?

Tail docking should be done before the lamb leaves the lambing pen, while it is easy to catch. You will need the help of an experienced adult the first time you do this.

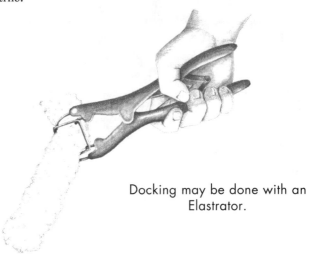

Docking may be done with an Elastrator.

Tails can be docked by several methods. Some people favor the Burdizzo Emasculator and knife. (This tool can also be used for castration.) Apply an antiseptic and, in warm weather, a fly repellent, to the wound. Some people prefer the Elastrator, which places a small, strong rubber ring around the tail. After the Elastrator band has been on for a couple of weeks, the tail drops off. Any method causes slight discomfort, but it is temporary.

What is castration, and is it necessary?

Castration is the removal of a ram's testicles, so that it is unable to breed. Your ram lambs should be castrated unless you plan to keep them or sell them for breeding. If you plan to sell the lambs at auction or to a packinghouse, you may be penalized a dollar or more if the lambs have not been castrated. The meat from older and heavier, uncastrated lambs can have a strong taste.

Castration. *The removal of the ram lamb's testicles so that it will not be able to breed.*

Male lambs are also castrated in order to prevent them from breeding with other lambs when they are penned together. Lambs can breed at a very early age. If your ewe lambs are bred too early, *serious* lambing problems can occur.

When and how do we castrate lambs?

Male lambs should be castrated early, but not in the lambing pen. Before the lamb can be castrated, both testicles must have dropped into the scrotum. This will take up to ten days. You should again seek the help of an experienced person until you learn the procedure.

You can use the Elastrator or the Burdizzo Emasculator for this procedure, too. Neither causes loss of blood, and there is little shock to the animal and only a slight risk of infection. With the Elastrator pliers, you stretch a very strong rubber band around the scrotum. The Burdizzo crushes the blood vessels to the testicles without breaking the skin.

Vaccinations

What vaccinations are needed?

Sheep require *vaccinations* (inoculations to protect them from certain diseases) just like other animals and people. Lambs should be vaccinated at an early age. Consult with your veterinarian, Extension Service agent, or experienced shepherd in your area about vaccinations for your sheep.

Sheep are commonly vaccinated against diseases that infect the lungs, the digestive system, and the reproductive tract. For example, lambs, as well as ewes and rams, can be vaccinated with NASALGEN-IP, an *intranasal* vaccine that helps to protect against respiratory disease (pneumonia). It should be given to lambs during the first three to seven days of age.

All lambs should receive immunizations against

Vaccination. An inoculation that prevents certain diseases.

enterotoxemia, a condition caused by overeating. Ewes should be vaccinated and/or given booster shots for Clostridial diseases *before* lambing. (Use this opportunity to vaccinate your ram, too. You will be less likely to forget any sheep if you do them all at once.) This not only protects the ewe, but also increases the disease-fighting antibodies the ewe passes on to her newborn lambs when they nurse. COVEXIN-8 covers all common Clostridial diseases that affect sheep. Another group of vaccines can be given to ewes to protect them from diseases that may cause them to lose their lambs before birth.

Disease-control programs are hard to understand until you gain experience with sheep. The important thing to remember is that *most of the vaccinations must be given either prior to breeding or prior to*

Health Maintenance Record

Name _____ Ear tag # _____

Birth date _____ Birth weight _____

Color _____ Sex _____

Breed _____ Dam and sire _____

Vaccinations		Worming	
Product	Date	Product	Date

Lambing history

Exposure date _____

Lambing date _____

Ear tag #'s of lambs _____

Misc. comments

Use this to help keep track of your lamb's progress.

lambing. Don't wait until the ewes are almost ready to lamb before you vaccinate them.

Worming

Do all sheep need to be wormed?

Sheep have a high resistance to disease but low resistance to internal parasites. All sheep need to be wormed, especially lambs. There are many different kinds of worms that sheep pick up throughout the year while grazing. The worms are very tiny, and you cannot see them without a microscope. These worms suck blood. Lambs can die from severe worm infection. Again, consult with your veterinarian, Extension Service, youth group leader, or an experienced shepherd if you have any questions about worming.

Is there any way to cut down on worms?

If you are just starting with sheep, you will have a clean pasture, not contaminated by previous use. By cross-fencing (see pages 23-24), you can divide a large area into several small pastures so that your sheep can be rotated from one section to another. This helps reduce the exposure to worms and gives the pasture a rest. Pasture rotation, combined with clean feeding facilities to avoid pollution of water and feed, can help prevent a rapid or serious build-up of stomach worms.

How is worming done?

Even with precautions, sheep can get infected with worms, and they will need to be treated periodically. There are several good worm medicines that are safe even for pregnant ewes and small lambs. Obtain worm medicine from your veterinarian so that you can get correct information about proper dosage and timing. You may use a large pill called a *bolus,* which is placed at the base of the sheep's tongue with a bolus "gun," available at farm supply stores. Worming medication also comes in other forms, such as liquid "drenches," pastes, feed blocks, and injections.

How do I know a lamb has worms?

Lambs with worms may be underweight for their age, have potbellies, prominent hip bones, a scruffy wool coat, runny or loose manure with manure build-up on their rear, and *anemia* (blood loss caused by blood-feeding worms) that makes them weak. Anemia may be prevented by worming lambs when they are two or three months old. If you have only one or two sheep on clean pasture, you may need to worm your lambs only once more during the first year. In large flocks with limited or contaminated pasture, the lambs may have to be wormed monthly. Adult sheep with heavy worm loads will sometimes have a swelling under their throat or chin called "bottle jaw." Ewes should be wormed before breeding, before lambing, and before going out on fresh spring pasture.

How should I treat my sheep after they have been wormed?

Ideally, you should put your sheep into a clean pasture 24 hours after worming. In this way, the sheep will expel the worms and eggs in the old pasture, and will not contaminate the new pasture as rapidly. This is possible only if you have arranged cross-fencing in the pasture.

External Parasites

How do I treat external parasites?

There are two major kinds of external *parasites* (insect pests) that get on sheep. They are sheep ticks and lice.

Both cause severe itching, which can make the sheep very uncomfortable. If untreated, these parasites cause loss of wool and damage to a potentially valuable pelt.

What are sheep ticks?

Sheep ticks are not really ticks at all. They are actually wingless flies called *keds,* which look like ticks. Keds can be easily controlled by insect-killing sprays, pour-ons, and dusts. You can get rid of keds completely if you treat them with an effective pesticide. You may have to apply the pesticide twice, the second application about three weeks after the first. Your flock will stay free of keds unless you bring in an infected animal or let your sheep come in contact with infected animals (such as at a show).

When should I treat my sheep?

The best time to treat sheep for keds is immediately after shearing (but wait until any shearing cuts have healed). Be sure to treat *all* the flock, including lambs and rams. Move the sheared wool at least 50 feet away from the flock, since any keds in it will crawl out in search of the animals.

USDA, AGRICULTURAL RESEARCH CENTER, PHILA.

As with any pesticide, *read and follow the directions* very carefully for proper dosage and method of application, and precautions regarding use on pregnant ewes or lambs, as well as on lambs that will be sold for meat. The products with the active ingredient permethrin (synthetic pyrethrum) are highly effective and safe.

Sheep keds and eggs

Lice

How can I tell if my sheep has lice?

Less common than keds, lice are external parasites that bite the sheep and cause severe itching. Lice are very

HOOF TRIMMING TOOLS

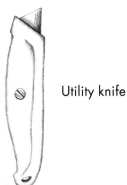

Utility knife

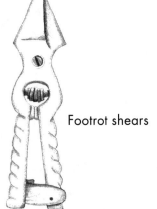

Footrot shears

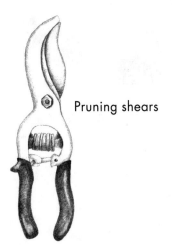

Pruning shears

small — so small that they can barely be seen with the naked eye. Two types of lice affect sheep — those that suck blood and those that bite. Both cause intense itching. Sheep with lice will constantly rub and scratch against fences, posts, feeders, and almost any other object around the farm. A farm infested with sheep lice will have bits of wool hanging on just about everything the sheep can rub against.

Treatment for lice is the same as for keds.

Hoof Care

Wild sheep and those grazed on mountainous pastures keep their hooves worn down by travelling over rocky ground. Domestic sheep walk primarily on soft soils. Consequently, their hooves become long and over-grown, and walking is painful. Sheep need to have their hooves trimmed so that they can walk properly, and also to help prevent hoof diseases such as *scald* and *footrot*. Before buying an adult sheep, be sure to look at its feet. Ask when they were last trimmed. If they show any need for trimming, the owner should be able to demonstrate the procedure for you. If the owner *can't* show you how, don't buy the sheep — this is a sign that the sheep's hooves have probably not been properly cared for.

How often should I trim my sheep's feet?

How often to trim depends on your pasture, paths, and barn floor. Many shepherds must trim twice a year, while others need to trim only once a year. If you notice limping sheep or any sheep on their knees, check their feet. When you are taking care of other needs, like worming or shearing, make a habit of trimming hooves, as well. Another good opportunity for hoof trimming is when you release ewes from the lambing pen.

How is trimming done?

For safety, wear leather gloves so you don't get cut by the hoof shears if the sheep becomes unruly. Set the sheep or lamb on its rump (as described on page 29). Trim the rear feet first, then the front feet. A kicking sheep, not held firmly or properly, could injure you with the sharp edges of the freshly trimmed hoof. (See illustration for guidance on trimming hooves.)

What do I use to trim hooves?

There are many styles of hoof shears on the market, but few are as easy and safe to use as Felco No. 2 pruning shears, made in Switzerland. Many farm and garden stores sell them. If you have many sheep to trim, the investment is well worth the money.

What is footrot?

Footrot is one of the worst diseases that can infect your sheep. It is a bacterial disease that causes hard-to-cure infections. Sometimes an animal that has no symptoms of the disease can carry the infection and spread it to others. This is called a carrier animal. Active outbreaks of footrot occur during times of warm, wet weather. The bottom of the hooves of infected sheep literally rot off, exposing the soft tissues beneath. This is an extremely painful disease. Regardless of how much you might want to buy a particular sheep from a footrot-infected flock, *don't*. Footrot can only be introduced into a clean flock by a carrier animal, and *you don't want to purchase an infected or carrier animal.*

Before buying any animal, look at the rest of the flock. If you see any limping sheep or sheep with severely overgrown, misshapen hooves, go somewhere else. Once footrot is introduced into your flock, it is almost impossible to get rid of it.

Some sheep raisers disinfect their sheep's feet after

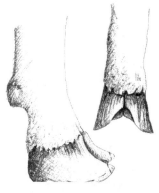

Overgrown hooves in need of trimming.

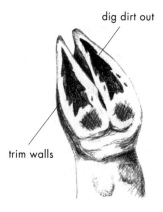

dig dirt out

trim walls

Trim hoof walls and remove dirt between toes.

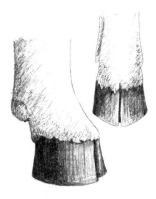

Properly trimmed hooves are similar to those of a half-grown lamb.

Possible Causes of Lameness Other Than Footrot

- Overgrown, un-trimmed hooves

- Wedge of mud, stone, or other matter stuck between the toes

- Plugged toe gland

- Abnormal foot development (inher-ited defect)

- Foot abscess

- Foot scald

- Thorns, punctures, stone bruises, etc.

they've taken them to shows, where they have been in contact with other animals. To disinfect, you must walk them through a footbath containing a special solution that kills the germs that cause footrot.

How do you treat a plugged toe gland?

Squeeze to remove plug, then disinfect with Listerine or hydrogen peroxide.

What is foot scald?

Foot scald is a skin infection that occurs between the toes. It is not caused by the same type of bacteria that cause footrot. The first sign of scald is limping. Usually, the front feet are the first ones that get sore. On sheep with foot scald, the skin between the toes is moist, hairless, and red (irritated). The skin can be rubbed away, leaving a raw, painful area. Scald infections are more common after a long, wet period. The infection can be treated with any of the disinfectants used to treat footrot. Move the affected animals to a clean, dry area.

Wool and Shearing

Shearing is usually done once a year, in the spring, so that the sheep are free of their heavy wool coats for the hot summer months. Shearing is very important for keeping your sheep healthy and comfortable, and this annual "haircut" has benefits for the shepherd, as well.

The first time you watch a shearing, you may find it quite a surprising scene — the sheep must be held in awkward positions while its thick wool coat, called the *fleece,* is removed with sharp clippers or shears. But the fact is that the sheep feel a lot better after shearing. It's also easier to keep shorn sheep free of parasites, and that keeps them more comfortable, too.

Aside from healthier, happier sheep, the main benefit of shearing for the shepherd is the wool obtained from each sheep.

Wool is a unique fiber. It is the *only* fiber in the world that retains its warmth when wet. This is why sailors of old always wore wool sweaters and clothing. If you do not spin your own wool, you can sell your fleeces to spinners who spin it into wool yarn for making sweaters, socks, blankets, and other wool goods.

Several characteristics determine the quality of a fleece, including wool type, strength, and cleanliness from vegetation. The type of wool — silky, curly, or

Fleece. *The thick wool coat of a sheep.*

coarse, for instance — is a characteristic of that breed. The strength of the wool is determined by health and nutrition. But even if your sheep has a strong fleece of a desirable type, it will be less valuable if it is not kept clean. To keep a fleece clean, you can use a sheep coat to protect it from vegetation such as seeds, hay, burrs, and bedding. (For directions on how to make sheep coats, see pages 81-83.) Eventually, you will earn a reputation among spinners for having prime-quality, clean fleeces, and your fleeces will be in demand far ahead of shearing time.

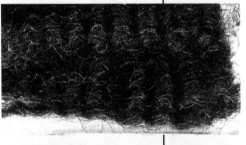

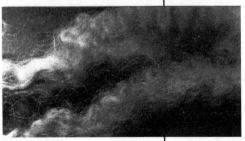

Fleeces from different breeds have different amounts of crimp.

Crimp

When you look closely at a handful of raw wool, you will see little zig-zag waves in the strands of wool. These are called the *crimp,* and they are what makes the wool elastic or stretchy. Wool with good crimp acts like a spring. When wool fibers are stretched, they spring back to their original shape. As a rule of thumb, the finer the wool strands or fibers, the finer the crimp. In very fine wool, the crimp can barely be seen. In very coarse wool, the crimp can vary from ½ to 3 inches between the "zig" and the "zag."

Tips on the Shearing Process

Sheep don't seem to enjoy the process of shearing, but we are sure that they are happy to be rid of their heavy winter coat of wool. To reduce the stress for both sheep and shearer, here are a few suggestions:

- Don't give your sheep food or water for about 12 hours (or overnight) before shearing. Your sheep will be more comfortable during shearing if it has an empty stomach.

- Be sure your sheep is dry. If rain is predicted, keep the sheep indoors the night before.

- Get your sheep into a pen where it can be caught easily. A shearer must not be expected to chase or catch your sheep.

- Prepare a good shearing floor. A smooth, plywood platform is better than a dirt floor or grassy area. It will be less slippery if a tarp or an old wool rug is spread on it. A tarp is more customary, but because it can bunch up and wrinkle, some people prefer using a wool rug.

- When shearing is done indoors, provide sufficient light.

- Have cold water or other beverage available for the shearer.

- Handle your sheep carefully, catching it properly (never by the wool).

- Have first-aid supplies on hand in case the sheep — or the shearer — is severely cut. Don't worry about little nicks on sheep. They may look bad, but the lanolin in sheep's wool helps them heal quickly. In hot weather, spray them with a fly repellent.

- Handle the fleece carefully, so that it doesn't get contaminated with manure, straw, or dirt. Sweep the floor after each sheep.

- Arrange a *skirting table*. This can be just a raised table with a slatted top. Sanded lath, long dowels, or 1" x 2" wood (on edge) makes a tabletop suitable for skirting. Throw the fleece on it cut side down to shake off any *second cuts* (see page 53). Skirt the edges to remove poor-quality or contaminated wool, and manure tags. (See illustration, page 52.)

Tags *are little tufts of wool that are so soiled and matted with manure or mud that they are solid. They should be removed from the fleece when skirting.*

Sheep-ish History

Did you know that one of the major causes for the American Revolutionary War was the restrictions England placed on sheep and wool?

During the colonial period of America, the King of England forbade the colonists to purchase wool-spinning machinery or sheep from Europe. He wanted to keep the colonists dependent on English-made woollen goods. This and other unfair laws angered the colonists so much that they decided they did not want to be ruled by the English king any longer.

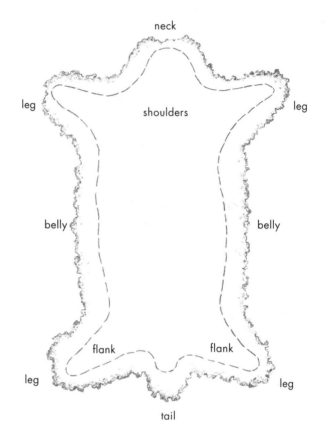

The dotted line shows where to skirt the fleece.

- Tie fleeces with *paper twine* (available from sheep supply stores) or place them in paper feed sacks (turned inside out) or cardboard boxes.

- Shearing is a good time to trim hooves, check udders, worm all the sheep, treat them for ticks and parasites (after the wool is removed), and check the teeth of oldsters.

Position the sheep for shearing by slipping your left thumb into its mouth behind the incisor teeth, and placing your other hand on its right hip.

Bend the sheep's head sharply over its right shoulder, swing the sheep toward you, and lower it to the ground.

The sheep sits resting against the shearer, as the shearer holds the sheep's head down and shears its left side.

Second Cuts

Second cuts are short lengths of wool created when the shearer runs the clipper over a place on the sheep that has already been sheared. This makes the sheep look smooth and lovely, but the short snips that result are very difficult to remove from the fleece when preparing it for handspinning. If they are not removed, they make unsightly little lumps in the yarn and weaken the yarn. When shearing for handspinners, it is important to avoid making second cuts.

Advantages of Hand Shearing

- No source of electricity is needed

- Less noise to annoy the sheep

- Blades are lightweight and easy to carry

- Easier to sharpen than power clipper blades; one needs only a hand stone

- Hand shears are less expensive

Shearing Your Own Sheep

Lots of kids have learned to shear their own sheep. Don't worry that your lack of size or weight will be a disadvantage when shearing. Some women shearers are no larger than many 12 year olds, and they do very well. When you get your sheep into the basic shearing positions, it will be helpless, because it won't be able to get leverage to scramble to its feet. Each of the shearing positions enables you to shear a different portion of the animal. Some of these are shown on page 53. Some people think these holds look like wrestling holds.

If you have more than one sheep, a great advantage to shearing your own is that you don't have to shear them all at the same time, as happens if you hire a shearer. If you have ewes with small lambs, shearing can be a very confusing and upsetting time. The sheep look and smell different after shearing, and all the lambs cry as they try to find their mothers. If you shear only one mother each day, this doesn't happen.

Another advantage of shearing your own sheep is that you gain familiarity with each animal. This is a good time to check the teeth, udders, and general health of each sheep. And, you will have time to handle the fleece in a way that is in keeping with its intended use. For instance, if you will be using the fleece yourself, you may want to sort and bag the parts separately: the belly wool, skirtings, and tags can be used for feltmaking; the prime fleece for spinning or selling to local handspinners. If the wool is going to be handspun, be especially careful to keep it clean of vegetation (burrs, seeds, and so on) and to avoid second cuts. It will be much more valuable.

Fitting and Showing

If you like adventure and excitement, you can find them in the show ring at your local, county, or state fair. You will need to know many things, and one of the best ways to learn them is to join a 4-H club. (State 4-H leaders are listed on pages 111-12.) Your 4-H club leaders and fellow members can help you with every phase of preparation and showing. Your skills at showmanship will come through training, practice, and show ring experience. But no amount of expert showmanship can make up for a poorly trained lamb.

Jamison White likes to show lambs, because you learn a lot, make friends, and maybe even win ribbons.

Fitting stand. *Used to hold the sheep still while it is groomed for a show, a fitting stand is a low platform equipped with a brace for the sheep's head.*

Equipment Needed for Fitting Your Lamb

- Mild detergent
- Hand shears
- Washing brush
- Rope halter
- Feed pan and water pail
- Blanket
- Large and small wool cards
- A "show" box to contain your equipment
- Hoof shears
- Fitting stand

Preparation

Find out the exact show requirements far enough ahead so that you will have time to comply. These could include such things as special rules regarding health papers, required vaccinations, shearing times, and so on. If you are going to several shows, you may find that the rules differ from one to the other.

Begin to assemble the necessary equipment well ahead of show time. For instance, build or purchase your *fitting stand*, and get the lamb used to it before it is needed.

Jaeme Lee Griffin, in her African wool outfit, won first prize with her Tunis sheep.

Training

Start training early — at least two months before the show — by using a halter (see page 28). Although a halter is not always used in the show ring, it is the best way to begin training your lamb. Also, you will want to use a halter around the fairgrounds and anytime you have the lamb away from home. You can learn the "tricks of the trade" from your 4-H leaders or others who are experienced at showing. Attend a show or two just as an observer before you enter one yourself.

Feeding

Make a gradual change to the feed your lamb will be getting at the show, if it is different from the lamb's normal feed. Make the change over a period of two weeks. Reverse this procedure after the show. You may also wish to bring water from home for use at the show in order to avoid any digestive upsets from a change in water.

Washing and Blocking

Lambs need to be washed prior to the show. Trim the feet before washing the lamb. Should your lamb's feet get dirty after washing, soak them in a can of soapy water to loosen dried mud and remove stains that may get on the wool.

Different shows have different rules about how short to shear, as well as how close to show time to shear the lamb. Find out well ahead of time, so that you can follow the proper procedure. After shearing, the lambs are combed with the wool cards and trimmed with hand shears until they are perfectly smooth. This is called *blocking*. The purpose of blocking is to improve the appearance of the show animal.

BLOCKING TOOLS

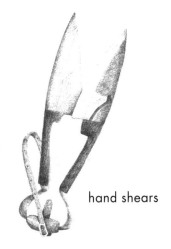

hand shears

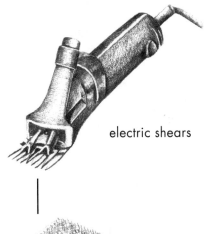

electric shears

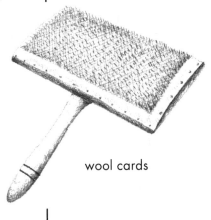

wool cards

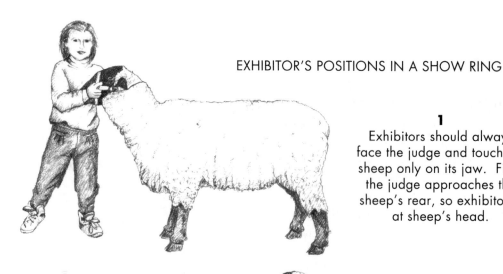

EXHIBITOR'S POSITIONS IN A SHOW RING

1
Exhibitors should always face the judge and touch the sheep only on its jaw. First, the judge approaches the sheep's rear, so exhibitor is at sheep's head.

2
The judge views the sheep from its right side and exhibitor moves to other side.

3
The judge now moves to the front of the sheep to look at its head.

4
Last, the judge moves around to look at the left side of the sheep.

Show Ring Etiquette

- Lambs should be clean and trimmed to meet specific requirements of the show.

- As an exhibitor, you should be well groomed and wear neat, clean clothes. Don't wear a hat or chew gum.

- Be familiar with the show schedule and be prompt.

- Practice polite show manners both inside and outside the ring.

- Don't pull the lamb's wool or touch the lamb except under the jaw, at the dock (rear end), or under the belly; you will also touch it when you set its legs.

- Assume a half-kneeling, squatting, or standing position by your lamb. Don't kneel in the sawdust; it will end up on your clothes and the lamb's fleece. (Rules on standing or kneeling may differ from state to state.)

- Don't crowd yourself or your lamb — you want the judge to see it. Leave 2 to 3 feet or more between you and the next exhibitor.

- Don't appear clumsy in the ring. Always cross over in front of the lamb, never behind it.

- Lead the lamb on your left side with your left hand under its jaw.

- Set your lamb up with its feet squarely placed, so that all four of its legs extend straight down from the "corners" of its body.

- Know your lamb's weight, breed, and age. Be prepared to tell the judge if you are asked.

- Be aware of the judge's position in the arena. Follow his or her instructions correctly and promptly. And *smile!*

- Pay attention to what's happening in the ring. Don't talk or look at anyone outside the ring.

- If your lamb doesn't cooperate, be patient, stay calm, and keep trying. Your lamb is probably nervous also!

- Be a humble winner and a cheerful loser.

Breeding and Lambing

The normal breeding season for sheep is in late summer and fall, from about mid-August to mid-November. Some areas have a breeding season in the late spring during the months of May and June. A ewe gives birth about 147–153 days (five months) after she is bred (becomes pregnant).

Special Feeding Requirements

Even before the ewe is bred, proper feeding plays a very important role in breeding. Before breeding, it is necessary to increase the energy value of the ewe's feed to be sure she is well nourished (*not* to fatten her). This process is called *flushing*. Flushing increases the number of eggs released by the ewe at breeding. The end result is increased fertility and more twinning.

You can flush your ewe by grazing her on a new lush pasture, by feeding her a small amount of grain, or both. About three weeks before you plan to breed the ewe, begin feeding her ¼ pound of grain a day. Gradually increase this amount to 1 pound per day. Feed 1 pound a day for seventeen days, then breed the ewe, and gradually reduce her grain. If the pasture or forage

Breeding. *The behavior, timing, and activity related to producing offspring.*

Estrus. *The time during which the female is in heat and can be bred. A ewe is usually in estrus for about 28 hours. If she is not bred in that time, her cycle continues, and she will come into estrus again in 17 days.*

Pregnancy. *The time required for the growth and development of the baby from conception to birth. Also called gestation. In sheep, it averages 147 to 153 days (about 5 months).*

Crotching. *Trimming the wool from around the dock (rear) area of the sheep.*

is of low quality (as it often is in most parts of the country at breeding time) continue a low-level grain feeding for three weeks after the ewe is bred.

If you own the ram, don't forget that his nutritional requirements need to be increased, also. Feed him just as you feed the ewes.

When the ewe is about 30 days away from lambing, increase the grain slowly to 1¼ to 2 pounds per day (depending on the size of the ewe). The ewe needs these additional calories for the growing lamb or lambs she carries. Failure to meet these nutritional demands may result in weak lambs, low birth weights, or worse — a very sick ewe that could die. Pay particular attention to the ewe during the last two weeks of pregnancy. If she acts weak or listless, call your veterinarian for help immediately. Don't forget to provide a constant supply of salt, a mineral/vitamin supplement, and fresh water.

A pregnant ewe, particularly one that is fat or sluggish, needs plenty of exercise, especially in the last month before lambing. If you feed her at some distance from the barn, she will have to move around in order to get something to eat.

Some producers breed young ewes to give birth at a year of age, while others hold them over for a year and breed them to lamb as 2 year olds. As a rule of thumb, ewe lambs should weigh 85 to 100 pounds (at least 65 percent of their mature weight) before they are bred.

Young ewes (those bred as lambs) need more feeding and management than older ewes. They will still be growing while pregnant and continue to grow as they are producing milk. Furthermore, their maternal instinct is not as well developed, so there is a greater chance that they might reject the lamb.

If your ewe has long fleece or a lot of manure around her rear end, you should *tag* or *crotch* her, by shearing the wool from the udder area and around the dock. This reduces the risk of a lamb sucking on a dirty tag of wool before it finds a teat.

The Breeding Process: The Ram

On your special sheep calendar, write the day the ram (sometimes called the *buck*) went in with the ewes, and any days that you observed breeding.

During breeding season, the rams may become very aggressive. *Do not enter the pasture or pens with the rams without close adult supervision.* The docile ram that came up to you all summer to get his back scratched can suddenly become aggressive and danger-ous, and may butt you in defense of "his" breeding group.

High daytime temperatures can cause rams to become infertile (unable to breed successfully). In very hot weather, confine the ram in a cool place during the day. He can be turned in with the ewes at night (if you are ready to have them bred). If hot weather infertility occurs, it will take about 45 days for the ram to become fertile again. Your veterinarian can obtain a sample of

Fertility. *The ability of the female to produce healthy eggs and become pregnant, and the ability of the male to produce healthy sperm.*

Infertility. *In a ram, the temporary loss of his ability to produce sperm or mate success-fully with a female. In a ewe, the inability to produce or release an egg or maintain the pregnancy.*

Norlaine Schultz's Perendale ram.

semen (the liquid that carries the sperm) from the ram to determine if hot weather has damaged the sperm. Even when the ram appears to be breeding normally, if the sperm is damaged, fewer ewes will become pregnant.

Preparing for the Arrival of Baby Lambs

While you are eagerly waiting for the birth of new baby lambs, there are several things you can do to get ready.

Lambing pens. Lambing pens are called *jugs*. A jug is a small pen about 4' x 6' square and about 36 inches high. Jugs can be made from a variety of materials, and they don't have to be elaborate. (If you would like to make one, see pages 87-90.)

Hand shears. These are used for *tagging* (removing excess wool from around the udder and under the tail). Be careful that you do not cut the udder or teats!

Newborn lambs are kept in jugs with their mother.

Antiseptic-lubricating ointment. You'll need these for your hands if you have to assist in a delivery.

Supply of paper towels and old, clean bath towels. You'll need these for drying the newborn lamb.

Hot-water bottle and/or hair dryer. These are for use with chilled lambs. If you think you may use the hair dryer, make sure you have an extension cord, too. (See pages 72-73 for how to recognize and treat chilled lambs.)

Strong iodine or Xenodine. This is used to disinfect the umbilical cord.

Clean water bucket. One of the first things a ewe will do following lambing is to take a big drink of water (and they prefer *warm* water).

Livestock molasses. You can put a little of this in the water for ewes that have just delivered a lamb. It gives them a boost of energy. You can buy this at the feed store.

Lamb bottle and nipples. A baby bottle and a nipple with a slightly enlarged hole is better for the newborn lamb than the standard lamb nipple, which is some-what larger. (Enlarge the hole by making a small cross with a sharp knife.) If you have a bummer (orphan) lamb to feed, you will need the standard lamb nipple once it reaches a week of age. When new, these tend to be a bit stiff for week-old lambs. If you place them in boiling water once or twice, they will be easier to use. It helps to vent the bottle if you lay a rubber band over the lip of the bottle before you fit the nipple on. With age and use, a nipple tends to become soft and loose. Avoid using an old nipple that the lamb can pull off the bottle and possibly swallow.

Frozen colostrum. If possible, have a few ounces of colostrum (see pages 69-70) milked from a heavy milking ewe or a ewe that has lost her lamb. Frozen

Purpose of the Jug

- Aids in *mothering up* (the bonding of the mother and lamb, a time when they learn to recognize each other's sound of voice and smell, and when the ewe accepts the lamb as her own)

- Protects the lamb from being trampled by other ewes in the barn

- Prevents drafts

- Keeps the lamb from getting lost

- Prevents the ewe from going outside and exposing her lamb to harsh elements (snow, rain, etc.)

- Provides you a better opportunity to observe your ewe and lamb during the critical first days of the lamb's life colostrum is great for use in emergencies, and can be kept in the freezer for a year or more and thawed when needed. *Do not thaw in the microwave.* This will destroy the antibodies that protect the lamb from disease. Always thaw colostrum slowly at room temperature or in warm water.

When Lambing Time Approaches

About a week or two before you think the ewe will lamb, start checking the development of her udder, which often gets larger as lambing time nears. Checking udder development is called *bagging the ewe*. *Note:* This is not always a dependable sign. Some ewes develop large udders up to three weeks before lambing, while others show practically no udder until after the lamb is born.

If your ewe goes off feed or shows any signs of illness close to lambing time, consult your veterinarian immediately. Sudden signs of illness prior to lambing could indicate a condition called "pregnancy toxemia." Signs of illness just before or after lambing could also be caused by a calcium deficiency. The signs of these conditions are similar and confusing. Either one should be considered an emergency. Call your veterinarian at once. He or she can tell the difference and treat the illness immediately.

Make sure the lambing area is safe: Look for holes among the hay bales that the lamb might climb into or heavy objects that might fall on it. Make sure heat lamps aren't put where they could fall and cause a fire.

Labor

At 24 hours before the birth of the lamb, you may see some signs that labor is about to begin.

Teats. *The fingerlike projections on the udder (bag) through which milk flows to the baby.*

Udder. *The organ that produces milk; also called the bag.*

Labor. *The process of giving birth, during which contractions push the baby out; for ewes, usually a period of one to three hours.*

Contractions. *Strong muscle movements in the mother's uterus that push the baby out at birth.*

First stage: *Dropping of the lamb (about 24 hours before birth).* The ewe's abdomen will appear to droop, producing noticeable triangular-shaped "hollows" just ahead of the hip bone.

Second stage: *Early labor (about 6–12 hours before birth).* The ewe may appear uncomfortable and begin to paw at the ground before lying down. In this stage she is restless and may stand up and lie down frequently. The ewe will be extremely unhappy if you attempt to pen her up at this time, because she wants to choose her own nest. Be patient; observe, but do not bother her. She may occasionally roll over on her side, grunt, and/or kick, momentarily rolling back into a normal resting position. She may go to many different locations in the barn or pasture before choosing her final birthing spot.

Third stage: *Real labor (about 1–3 hours before birth).* The ewe will act more uncomfortable and may get up and lie down more frequently. As the contractions begin, she may be observed "rolling" a bit and paddling slightly with her feet (as though she were swimming). As the contractions increase, she will often point her nose in the air and let out a grunt. Shortly after that, the reddish pink water bag will emerge and soon break. Don't worry if the ewe gets up and walks around with the water bag hanging out. She is just trying to find a more comfortable position. Shortly, the lamb will begin to appear. If the lamb hasn't started to emerge after 30 minutes or more of hard labor (indicated by heavy straining and grunting), get help!

In normal births, the lamb's front feet and head appear first. There are a number of abnormal lambing positions, and these can be further complicated if there are twins or triplets. Nearly all basic sheep books have drawings of abnormal delivery positions, as well as instructions on how to assist in all kinds of deliveries. Because of their small hands and arms, young people

If you think your ewe is having trouble, call for experienced help.

Uterus. *A hollow organ with muscular walls where the fertilized egg attaches, develops, and grows during pregnancy. Sheep have a two-sided uterus.*

Umbilical cord. *A long cord that connects the mother and the developing baby in the uterus. During pregnancy, it transports nutrients from the mother to the baby, and wastes from the baby to the mother.*

This cross-section view of a lamb in its mother's uterus shows a lamb in the normal position for birth.

can often be helpful at difficult deliveries. If you assist at a delivery, be sure to wash your hands first with hot, soapy water. You can also wear disposable plastic gloves when assisting. With a little practice and instruction from an experienced shepherd, you can learn to handle many of your own lambings and be a valuable assistant for neighboring sheep raisers.

If you are not sure how to assist the ewe, quit early and call for help. Nine out of ten deliveries happen without problems.

A two-hour-old lamb at Caretaker Farm in Williamstown, MA.

"When the lambs arrive,
You give them a rub;
Then it's time
To snip, dip, strip,
and jug."

After the Lamb Is Born

The first 10 or 20 minutes after the birth are critical for the new lamb, especially if the weather is extremely cold. There are three things you can do to help:

- See that there are no membranes covering the lamb's face or nose.

- Make sure the lamb is breathing. If it is not, rub it briskly with dry towels to stimulate it to breathe.

- Place the lamb at the ewe's head and allow her to clean off the lamb.

Try not to disturb the ewe for at least 30 minutes. If it has been a difficult lambing, she may be tired, and you will need to dry the lamb with a clean towel. Return it to the ewe and allow her to become attached to it. Too much activity or interference by you may cause the ewe to reject the lamb. Be patient and let nature work!

Once the ewe has obviously accepted the lamb (by licking it and making low, soft sounds), it is time to put the ewe and lamb in the jug and "snip, dip, and strip":

Jug. Place the ewe and lamb(s) in the jug. Carry the lamb slowly toward the jug. The ewe will usually follow, as long as she can see the lamb or hear it call out. (Under normal conditions, three days is plenty of time to keep the ewe and lamb in the jug. Ewes with very small lambs, twins, or triplets may benefit from an extra day or two of jug time.)

Snip. Use a scissors to snip off the umbilical cord about 2 inches from the lamb's body.

Dip. Fill a small, wide-mouth pill bottle half full of strong (7 percent) iodine or Xenodine. Hold the bottle tight against the abdomen and dip the umbilical cord in the iodine until the cord is saturated.

Strip. The teats have a waxy plug that must be removed before the lamb can nurse easily. Strip milk from both the ewe's teats, gently forcing out the wax plug. One squirt of milk from each teat will assure you that the milk can flow freely.

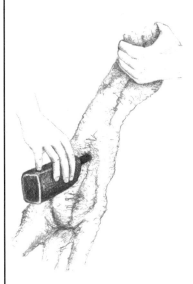

Snip the umbilical cord and dip the stub in a bottle of iodine held tight against its abdomen.

Gently strip the waxy plug out of the ewe's teat.

Nursing and Colostrum

Make sure the lamb is nursing. The normal lamb will be able to stand about 20 to 30 minutes after birth. As soon as it begins to walk, it will be hungry. Most lambs (with experienced mothers) will find the teats and begin nursing naturally within the first hour. If the

Colostrum. *The first milk after birth. Formed in the last weeks of pregnancy, it is rich in fat, protein, and protective anti-bodies.*

If your ewe has more than enough colostrum, collect and freeze some of it for emergency use.

lamb appears to have difficulty because it is weak or chilled, or has an uncooperative mother, it will need help. Twins or triplets need special attention. First, try to get the lamb to nurse naturally. If the lamb is weak or chilled, milk the ewe and feed about 2 ounces of the *colostrum* (the first milk) to the lamb with a bottle. Colostrum has antibodies to disease, as well as laxatives, minerals, and sugars to give the lamb needed energy.

Feeding the Ewe

Shortly after the lamb arrives, the ewe will be hungry. Feed her a good quality hay and a bucket of fresh, warm water, to which you've added 1 or 2 tablespoons of molasses.

Shortly after lambing, the udder in the normal ewe will fill with milk. If she is a good milking ewe, she will produce more milk than the lamb can drink, and the udder will become over-full. If you feed her extra grain, the additional protein it provides will be converted to more milk, which could result in a grossly over-full, painful udder. To avoid an excess of milk, do not feed grain for approximately three days after lambing. You should then begin feeding grain, gradually increasing it until you are offering 1½ to 2 pounds daily. Feed the grain twice a day, so she doesn't overeat.

Ear Tagging

If you have more than two or three ewes, you may want to identify their lambs by applying ear tags. Some are self-clinching; others require a small hole (just like piercing your own ears). The small Hasco-brand (or similar) metal ear tags can be applied whenever it is convenient for you. Some shepherds tag lambs shortly after birth (while they are doing the "Snip, Dip, and

70 Y o u r   S h e e p

Strip" procedure). Others tag lambs just before releasing them from the jug. *Record* the tag number, ewe number, and lamb's birth weight in your record book.

Ear tags are small, metal, numbered clips.

Docking and Castration

All lambs should have their tails docked (shortened) at an early age. Two or three days old is ideal, as it causes less pain and trauma than when they get older. Castration can be done at about 10 days. To find out how to do these procedures, see pages 39-41.

Problems with Newborn Lambs

Most lambs grow up without problems. But sometimes there are problems, and the ewe and her lamb or lambs will need help from you. Here is a list of the most common problems that can happen after birthing, and tips on how to handle them.

Ewes with No Milk

Sometimes ewes seem to have no milk. It might be because it is taking a little longer for their udders to fill with milk (called "dropping" or "let-down"), or they may have an infection of the udder called *mastitis*. If the udder is hot or lumpy, mastitis may be the problem, and you should call your veterinarian.

If the udder seems normal (soft and full), keep milking the teats every 15 minutes or so to see if the milk is beginning to flow. It sometimes take 2 to 3 hours before the milk lets down. In the meantime, the lamb must be fed colostrum. If you have another ewe that has just lambed, milk 4–6 ounces of colostrum from her and feed it to the lamb. If you have frozen colostrum on hand, now is the time to use it. Remember: *Do not thaw colostrum in the microwave.* You will destroy the antibodies that prevent disease. You can give the lamb an additional feeding of diluted canned milk (available in the grocery store) or lamb milk replacer 2 or 3 hours later. (Don't use milk replacer sold for other animals.)

If the ewe still does not have milk after 3 hours, an injection of a drug to stimulate milk let-down can be given. Keep the lamb with the ewe, but begin feeding it as if it were an orphan. (See pages 34-36 for how to feed orphan lambs.)

A Chilled Lamb

A chilled lamb will appear dull, listless, and weak. It may not be able to stand or suck. A slightly chilled lamb (temperature of 100°–102°F) may be warmed by placing a hot-water bottle against it or using an electric hair dryer on it. Check the lamb's temperature rectally with a glass-mercury or digital thermometer. If its rectal temperature is less than 100°F, the lamb needs immediate attention. (Another indication that the lamb is severely chilled is that the inside of its mouth will be cool when you feel it with your finger.) Immerse it up to its neck in warm water, comfortable to the touch, then gradually heat the water to about 110°–

115°F. When the lamb's body temperature reaches 100°F, its mouth and tongue will again feel warm. This may take several hours. Keep it in the warm water until its temperature is 102°F (normal lamb temperature is 103.1°F). When the lamb is completely warm, rub it with a towel and dry it thoroughly with the hair dryer. Try bottle-feeding the lamb. Return it to the mother as quickly as possible and make sure that she will accept it.

Keep the towels used to dry the lamb. They contain the birth fluids that help the mother identify her newborn. You may need to rub the lamb again with the towels to put its scent back.

Baby lambs can tolerate cold temperatures very well, but only after they are completely dry and have received colostrum. In order to keep a weak lamb from becoming chilled in very cold or wet weather, wrap it in a homemade plastic or fabric lamb coat just before you turn the ewe and lamb out of the jug. If you'd like to make lamb coats, you'll find directions on pages 83.

If you are using a hot-water bottle to warm a chilled lamb, don't have it so hot that it burns! Heat applied to the belly is most effective.

When the Ewe Rejects Her Lamb

The rejected twin. It is fairly common for the ewe to reject one of her twins. The ewe may be nervous, inexperienced, or confused and accidentally lose her lamb's scent. You can work to graft it back to her by forcing her to accept it, bottle feed it, or "graft" it onto a ewe with more milk or one who has just lost her lamb. If the true mother is a young ewe or a very old one, she may not have enough milk for twins, and it's therefore best to relieve her of the rejected one. Some ewes with inadequate milk will attempt to raise twins, with the result that both twins suffer from lack of milk. One solution to this problem is to partially "bum" them by supplementing the ewe's milk with milk replacer.

The forgotten twin. Occasionally, when a ewe delivers the first of a twin, the bonding period will be interrupted by the arrival of the second lamb, which is dropped several feet away. Similarly, the bonding of second or third lambs is sometimes interrupted when the first lamb comes around to be nursed. If the ewe is distracted, she may totally forget about the other lamb. Be sure to check the entire lambing area. You may be surprised to find that the single lamb with the ewe has a brother or sister on the other side of the barn! Similarly, some ewes with twins are happy as long as they have contact with one lamb or the other and don't miss a lamb that wanders off or gets left behind. In this case, keep the ewe and lambs in the jug a few extra days until the lambs are strong enough to keep up with the ewe by themselves.

Difficult labor. This is one of the most frequent causes of rejected lambs. The ewe is so tired after labor that she simply doesn't care! These ewes often recover their mothering instinct after several hours of rest. Keep mother and lamb in the jug, but bottle feed the lamb.

Painful udder. A good milking ewe may have an over-full udder at lambing. This hurts, and the ewe is reluctant to allow the lamb to nurse. If this happens, milk out the udder until the excess pressure is relieved. Bottle feed a few ounces to the lamb, and freeze the rest for emergency use.

Lamb with sharp teeth. Check the lamb's mouth. If it has sharp teeth, use an emery board to file them down a little.

The chilled lamb. The ewe may abandon a lamb because she thinks it is dead. When treating a weak or chilled lamb, don't keep it away from its mother too long, or she may reject it when you bring it back (especially if you have had it in hot water, which washes off much of the lamb's natural scent).

Strong medicine odors. Be careful when treating the newborn lamb's navel with iodine. Iodine has a very strong, repulsive odor that sheep don't like. If you use too much, the ewe may reject the lamb.

The wandering lamb. A ewe may be very tired after delivery of a large, robust lamb. While the ewe is resting, a strong lamb will sometimes just get up and wander off in search of mama before the ewe has had a chance to bond to it. Getting them into the jug promptly will prevent this problem.

The stolen lamb. On rare occasions, a ewe will give birth to a lamb only to have another ewe — usually an older ewe who is close to lambing herself — come by and adopt it. Sometimes the "foster" mother and lamb will develop a very strong bond. The only problem with this is that the foster mother will have no milk left for her own lambs.

Trading lambs. This occurs fairly frequently when there are too many ewes and lambs in one group, or when jugs are not used. The lamb from one ewe will bond to another and vice versa. Everyone seems happy and no damage is done, except that the ear tag numbers don't match and it could confuse your record keeping.

Grafting

Grafting is the process of bonding an orphan lamb to another ewe. Sounds simple, doesn't it? Sometimes it is, and sometimes it isn't! The bonding between lamb and mother is based on the smell and "voice" of the lamb, but *not its appearance.* All sheep may look alike to other sheep, but they all smell different.

Research has shown that grafting is always more successful if done in the first 20 hours after the ewe has lambed. The orphan to be grafted should not be more

Grafting. *The process of bonding an orphan lamb to another ewe.*

than one day old. The sooner after lambing, the better.

Don't try to graft an older lamb on a ewe who has lost a twin, or who has an excess of milk and only a single, small lamb. An older lamb could consume more than its share of milk, causing the younger lamb to suffer from starvation.

Tricks to Make Grafting Work

■ Rub both lambs with molasses water, so the ewe will lick and accept them both.

■ Sometimes a little bit of vanilla or a nonscented room deodorant sprayed on a cloth and rubbed on the ewe's nose and lamb's rear will suffice.

■ If a ewe has just dropped a single lamb, take some of the watery fluid from the birth sac of that lamb and smear it all over the orphan you are attempting to graft to her. If the orphan is young and this is done within a few minutes of delivery, acceptance of the orphan lamb is almost assured.

■ Use an "adoption" or "fostering" coat. A fostering coat is another method of transferring a familiar lamb smell to a new baby. To use the coat, which is like a cotton tube sock, put it on the ewe's single live (or dead), wet lamb. The coat absorbs the lamb's smell within an hour. Then, turn the coat inside out and stretch it over the lamb you want to graft to the ewe. You can make your own fostering coat with any stretchy material, such as sweatshirt sleeves.

■ Feed the rejected lamb colostrum while you wait for the accepted lamb to have its first bowel movement. This will be a dark black, tarry material. Take a little bit of this and smear it on the rear of the rejected lamb. Now both lambs will smell the same, and if all goes well, the mother won't notice the difference.

■ The final solution, which has a high success rate, is to place the ewe in a *stanchion* — a piece of equipment that fits loosely around the ewe's neck and keeps her from moving forward or backward. She will have freedom to eat hay and drink water, but cannot see, smell, or butt the lamb. The lamb to be grafted is put in the pen with her where it can nurse as desired. You may need to hold the ewe for the first few times, so you know that the lamb has had its colostrum (and the lamb knows where to find the milk). It may take up to five days for a stubborn ewe to completely accept the lamb.

A stanchion holds a ewe still so her lamb can nurse.

Caring for the Ewe after Weaning

The ewe's milk production reaches a peak at about four weeks after lambing, then begins to decline. (For information about weaning, see pages 37-38.) If weaning occurs abruptly, the ewe can develop udder problems. To avoid this, reduce the ewe's grain ration about three weeks before weaning and eliminate it altogether about two weeks before weaning. The quality of the hay should be reduced about three weeks before weaning; this will also help cut the ewe's milk production. If the ewe is still *making a full bag* (if her udder is still filling with milk), it can be partly milked out once a day for a few days to reduce the chance for mastitis (udder infection). Milking more often would just stimulate continued milk production. Keep the ewe in a clean, dry area for a few days after weaning until the teats close and the udder stops swelling.

Baby lambs from Wool and Feathers farm in Vermont.

Sheep Projects

CHAPTER 10

Ways to Earn Money

If you use your imagination, you will find many ways of making money from your sheep. The money you earn can help with the costs of taking care of your animals, and perhaps you can even earn a little extra. In addition to the following, you can probably think up a lot more things to do.

- If you are interested in spinning, knitting, weaving, or felting, you may be able to sell the articles you make. Handmade objects also make the most appreciated gifts. Once people become aware of your skill, you may get orders for your handwork.

- Consider the needs of other sheep owners. Would they like to buy homemade lamb adoption coats? This would be a good Christmas present offer, since lambing season is soon to come.

- Money saved is like money earned. Lamb and mutton sausage and sheep-milk cheese save grocery money as well as use your products to advantage.

- If you slaughter your lambs, save the fat and make tallow candles. For instructions, see Phyllis

Hobson's *Making Homemade Soaps and Candles* (Garden Way Publishing, 1974).

- Sheep manure may seem worthless, but not to a gardener! Think of a clever name for your bags of sheep manure, package them with care, and offer them to gardeners in your area. Sheep manure is dry, has little odor, will not burn plants, and is an excellent substitute for commercial fertilizer. The "sheep-brand" fertilizer has more nitrogen, more phosphorus, and more potassium (valuable nutrients in fertilizer) than manure of cows or horses.

- Do you know how to shear? Custom shearing can bring in extra income. Professional shearers often do not want to bother with only two or three sheep at a stop. When the owner is a spinner, speed is not as important as a good, careful job. If the owner does not want the wool, then you can discount the shearing fee and keep the wool for your own use.

- How about "babysitting" for other sheep owners? You could make it known that you would be available to feed and water their sheep over a weekend or vacation. Get exact instructions about the time, quantity, and location of feed, as well as any special needs.

- If you are handy with wood, you can make sheep "furniture," such as lambing pens (pages 87-90) and feed troughs (pages 92-94). To gain experience, make these for your own use first. You will also have an idea about the cost of materials, and how much you should charge for them.

- One way to make money from fleece is to sell it to a *wool pool,* which combines fleece from many sources and then processes it. However, if you sell good handspinning fleeces directly to individual spinners, you can make considerably more money. To qualify, the fleece must be clean of debris (seeds, straw, etc.), sheared carefully, and skirted heavily. It should be from a good wool breed, with a *staple length* (the length of each wool fiber) of at least 3 inches. White wool should be exceptionally white and have no dark fibers. Brown, gray, or black wool can be of any shade, or white with an even sprinkling of dark fibers.

■ Provide a custom *scouring* (washing) service for your spinner customers. Careful scouring of a handspinning fleece could double its value, and this process does not require expensive equipment. See washing directions on pages 94-95.

■ If you have a drum carding machine, you can provide a carding service to other spinners. For free instructions on constructing a carding slope, write to Patrick Green Carders Ltd. 48793 Chilliwack Lake Rd., Sardis, BC V2R 2P1, Canada.

Things for Your Sheep to Wear

Sheep Coats

Sheep coats (also called blankets, covers, or rugs) can be used prior to showing, as well as all year round, to keep fleeces clean. In areas with severe winters, coats help the sheep conserve energy. The energy that might have gone toward keeping the sheep warm goes to increased wool growth and heavier lambs. Many owners of small flocks who depend on the sale of choice fleeces now use coats on all their sheep.

If you want to keep costs down, you can make fine sheep coats from woven-plastic feed sacks, which allow air circulation and hold up rather well. The edges must be well hemmed so that they don't fray. The easiest material to work with, however, is #10 cotton duck. Prewash it before constructing the coat to prevent shrinkage.

Staple length. The length of each wool fiber.

Scouring. Washing wool fleece in preparation for carding and spinning.

Use the dimensions suggested or measure your sheep and make a custom-fitted garment. For best protection, the coat could extend 3 inches below the belly. If your material isn't wide enough, make a seam along the center (at the backbone). If you are custom-fitting, determine the coat length by measuring from the center of your sheep's breast to the end of the back thigh. Make the hind leg loops of soft "pajama" elastic, which is about 1 inch wide and not as stiff as most elastic. Although the garment itself should last more than one season, the elastic will probably have to be replaced for the next season. Check the sheep often to be sure the elastic isn't rubbing its leg and causing injury.

To be sure you get the right fit, first make a rough model out of an old sheet and try it on your sheep. The front edge should be as close to the head as possible for maximum protection from hay and weed seeds. If you make a center back seam, you can get a closer fit by shaping the seam to conform to the curve of the sheep's back.

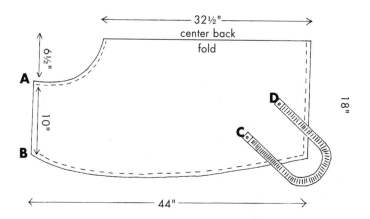

1. Match seam **A-B** with seam **A-B** on the other side, and sew, overlapping ½ to 3 inches, depending on the size of the sheep.

2. Make a ½-inch hem on outside edges (shown by dotted line)

3. At points **C** and **D** on both sides, stitch soft elastic, 24 to 27 inches long, depending on size of sheep.

Newborn Lamb Coats

In extremely cold and/or wet weather, coats can help the baby lamb conserve energy and prevent *hypothermia* (severe chilling). Put the coat on after the lamb is dried off or before you turn it out of the jug, depending on the weather. A plastic coat not only conserves body heat but also serves as a "raincoat."

Pajama elastic is soft and will not chafe the lamb if fitted properly. Because lambs grow so fast, you may want to make these in several sizes.

Make lamb coats in various fabrics to see what sells best:

- Lightweight canvas or cotton duck makes a practical garment.

- Old blue jeans fabric is already fashionably faded. Add a cotton-plaid lining for a western-jacket look.

- Ewe lambs would look cute in flowered prints or ruffles.

- Plastic raincoats can be cut out of plastic utility bags.

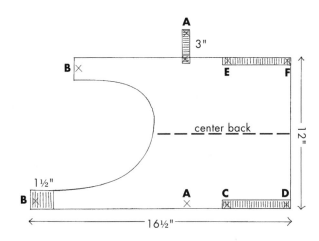

1. Belly band: stitch elastic to point **A** on opposite edge.

2. Neck fastening: stitch elastic to point **B** on opposite edge.

3. Leg fastening: stitch the ends of the 4-inch pieces of elastic on the inside of the coat at points **C**, **D**, **E**, and **F**.

Materials and Tools

❑ 6 (or 8–10 for training) feet of ⅜-inch 3-ply nylon rope

❑ Candle or other flame (to melt ends)

Rope Halter

A low-cost halter can easily be constructed from nylon rope. *Note:* For a show halter, which has a short lead strap, you will need a 6-foot length of ⅜-inch diameter 3-ply rope; if the halter is to be used for training and tying up, you will need a piece 8–10 feet long.

Caution: Ask an adult to help you with the first step.

1. Cut the rope by holding it over a flame at the desired length. Slowly rotate rope over flame. When the nylon rope has melted apart, and while the melted nylon is still hot enough to stick together, squeeze the ends in order to seal all nylon strands together. Use pliers or a similar tool to do this. *The ends of the rope will be very hot.*

2. Loop the rope together as shown in the drawings. The holes in the rope are made by twisting the ply open.

Note that the plied rope may be twisted apart to make a hole into which the end of the rope can be threaded.

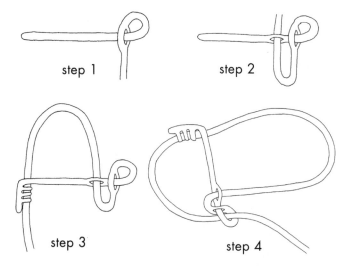

step 1
step 2
step 3
step 4

Cheesemaking

Most people have, at one time or another, eaten sheep cheese — even if they didn't know it. The best European gourmet cheeses — such as Roquefort, Romano, and Peccorino — are most often made from sheep milk. Sheep milk is ideally suited for cheesemaking. It contains almost double the solids of cow's milk and is high in proteins and minerals, so you can produce more cheese with less milk. It also contains a higher percentage of butterfat than cow's milk.

Collecting enough sheep milk for a cheese project could take quite some time for one person, but it is a great group project. You can collect, chill, and freeze the milk until you have enough to make cheese. With a lot of people helping, it shouldn't take long.

Cheesemaking does not require a lot of special equipment (see list). Many items you will already have in your kitchen, and you can easily make a cheese knife or press.

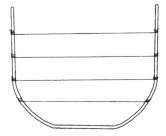

Cheese knife. Bend a firm wire into a U-shape that will fit into the kettle in which you will make your cheese. At intervals of 2 inches, anchor rustproof wire from one arm of the U to the other by twisting firmly.

Cheese press. Bore ½-inch holes through an 8-inch square, hardwood board at 1-inch intervals in both directions (checkerboard fashion). Sand the board, rub it with cooking oil, and wipe it clean. To use, put the cheese bag on this board and top it with another board the same size. On the top board, place a weight, such as a gallon jug of water. (C-clamps can be used instead of the weight.)

"Your Sheep" Cheese

This recipe makes a highly versatile, low-fat cream cheese. Use it as you would any cream cheese in your favorite recipe. It makes a great dip or spread when seasoned with parsley, chopped onion, pressed garlic, pepper, powdered soup or salad mixes, or herbs. Sweetened, it makes a delicious filling for cake.

> *1 gallon pasteurized, whole sheep milk*
> *½ rennet tablet*
> *¼ cup cold water*
> *½ cup fresh commercial buttermilk*
> *1–1½ teaspoons salt*

1. Pasteurize the sheep milk by heating it in a stainless steel kettle to 155°F and keeping it at that temperature for 30 minutes. Cool the milk to exactly 85°F.

2. Dissolve the rennet tablet in the water.

3. Add rennet tablet mixture and the buttermilk to the cooled sheep milk. Stir very gently for about 10 minutes or longer. Stop stirring the moment you notice a slight thickening or setting. If you stir too long, you will get a mushy product instead of a firm curd.

4. Keep the mixture warm — at 80°–85°F. Don't let it get any hotter or the rennet will be destroyed. Use and watch your thermometer. The best way to hold this temperature is to set your cheese kettle in a

Cheesemaking Equipment

- ❏ 6-quart, stainless steel kettle (*not* aluminum)

- ❏ Dairy thermometer

- ❏ Cheese knife (see page 85)

- ❏ Fine muslin bag
  or
- ❏ Cheesecloth and purée sieve or fine colander

- ❏ Cheese press (see page 85)

- ❏ Dishpan with rack

large pan of warm water to which you can add hot water from time to time as it cools. Let it stand until *whey* (a watery-looking liquid) covers the surface and the curd breaks clean from the side of the kettle (like Jell-o) when it is tipped.

5. Cut the curd into 1-inch cubes by running a long, thin knife through it in both directions (right to the bottom of the pot). Then cut the strips horizontally (as nearly as possible) by inserting your cheese knife and drawing it across the kettle.

6. Pour or ladle the mixture into a clean muslin bag or fine colander lined with cheesecloth. Allow it to drain (catching and saving the whey), until nearly all the liquid is gone. Press out the rest of the whey with your cheese press. If you don't have a cheese press, cover the bag with a dish and weight it down with a jar filled with water.

7. Keep the whey in the refrigerator until the cream rises and becomes firm enough to skim off. The cream will be of butter-like consistency. Work it back into the cheese, mixing thoroughly. (Save the thin whey to use as the liquid in bread baking, or feed back to the sheep.)

8. Once the cheese feels firm, work in the 1–1½ teaspoons of salt. Congratulations! You have just made creamy sheep cheese.

Over 36 million pounds of European sheep's milk cheese is imported into the U.S. each year.

Pens and Feeding Equipment

Lambing Pens (Jugs)

The following plans originated with the Norseman Sheep Co. many years ago, and the design remains unchanged. In fact, ten of the original jug panels are still in service. Although they cost slightly more to

Materials for One Jug

❏ Two 4' x 8' sheets of ½-inch exterior sheathing grade plywood, for end and side panels

❏ 2" x 2" x 74' lumber, for frames

❏ Nails (16-penny size)

❏ Screws (1½-inch drywall)

❏ Twine or wire to fasten the panels together

construct than those built of scrap lumber and old pallets, their durability makes them more economical over time. Because these jugs are solid at floor level, they prevent drafts. In addition, the ewe can readily look out over the 2-foot-high solid panel, so she doesn't feel trapped or confined.

The jugs measure 4' x 6' and are 3 feet high. They consist of end panels 4 feet wide and side panels 6 feet long. They have holes at the four corners where they can be laced together with wire or twine. When more space is required (such as for triplets), four side panels can be laced together to produce a 6' x 6' jug. Conversely, if you wish to confine a single lamb for some reason (sickness or injury, for example), four end panels can be used to make a 4' x 4' jug.

The jugs are set up side by side, and as many can be placed in a row as space permits.

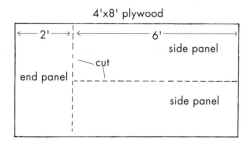

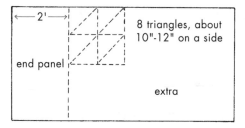

1. Cut a 2' x 4' piece from the end of each sheet of plywood. Cut one of the remaining sheets of plywood down the center to get two pieces that are each 2' x 6' long.

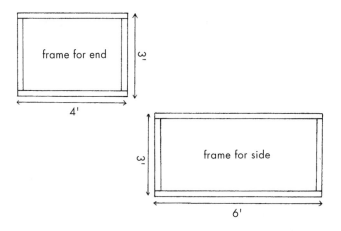

frame for end

3'

4'

frame for side

3'

6'

2. With the 2x2s, construct two frames for the ends, each measuring 3' x 4'. Use a single 16-penny nail to fasten the frames together at each corner. The plywood sides and corner braces that you will be adding in Steps 4 and 5 will strengthen them.

3. In the same way, construct two frames for the side panels, each measuring 3' x 6'.

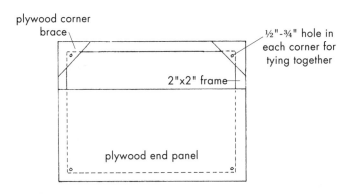

plywood corner brace

½"-¾" hole in each corner for tying together

2"x2" frame

plywood end panel

4. Screw the plywood panels to the frames, fastening the 2' x 4' pieces to the smaller frames, and the 2' x 6' pieces to the larger frames.

5. Cut eight triangular pieces of plywood (about 10"–12" on a side). Screw these to the top corners of the panels.

Features of a Good Jug

- Inexpensive to construct

- Solid sides to prevent drafts

- Easily set up or moved

- Adequate space for single or twins

- Folding or stacking for easy storage

- Durable for many years of use

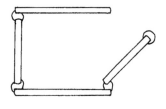

finished jug (top view)

6. Drill ½–¾-inch holes through the plywood at each corner. Run wire or twine through the holes so that you can fasten the panels toether at the corners. Use one end panel as the door. Tie it on one side, so that is can swing open. Put a loop of twine around the free side to keep it closed.

Creep Panel

A creep panel is a gate with expanded openings that allows the lambs to get through, yet keeps the ewes out. A small ewe can wiggle through a short and wide, or tall and narrow opening, especially when there is grain on the other side! The trick to an effective creep gate is to arrive at the perfect combination of height and width. The magic dimensions for most breeds are 8 inches wide and 15 inches high — no more and no less.

Creep gates can be made from a variety of materials ranging from welded pipe to 1" x 4" lumber (as shown). Creep gates take a lot of abuse. If you construct one from wood, *do not use nails; fasten all boards with bolts.* Fat lambs and big ewes trying to get in will wreck a nailed-together creep gate very quickly.

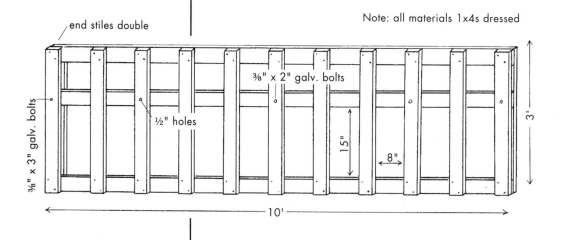

end stiles double

Note: all materials 1x4s dressed

⅜" x 2" galv. bolts

½" holes

⅜" x 3" galv. bolts

15"

8"

3'

10'

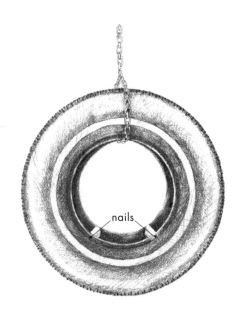

nails

Salt or Mineral/Vitamin Feeder

The best salt or mineral feeder money can buy can be made from an old, discarded passenger car tire. It will never rust, break, or tip over. These salt and mineral feeders have been used for years at the Norseman Sheep Co., where the first one made is still in service.

1. Insert the 2" x 2" boards between the beads of the tire at the 5 o'clock and 7 o'clock positions. (You may need assistance of an adult, because the beads of the tire can be difficult to stretch apart.) This will keep the tire spread apart so that when you place the salt/mineral mixture inside, the sheep can easily reach it.

2. Fasten the boards in place by driving a nail through the sidewall of the tire into the ends of the board.

3. Hang the tire with a length of chain from an overhead support in the shed or barn, so that it is free-swinging and suspended approximately 14–18 inches off the floor (depending on height of sheep).

Materials

- ❏ 1 discarded car tire

- ❏ 2 pieces of 2" x 2" lumber, 8 inches long

- ❏ Four 1½-inch galvanized roofing nails

- ❏ 1 length of light-weight chain, as long as necessary to reach from an overhead support in your shed or barn to about 14–18 inches from the floor.

Materials

- ❏ ½-inch waterproof sheathing plywood, 16" wide and 8' long

- ❏ One 2" x 4" x 6' fir (ends of trough and leg braces)

- ❏ Two 2" x 4" x 8' fir (sides of trough)

- ❏ One 2" x 4" x 12' fir (support legs)

- ❏ One 1" x 6" x 8' #2 pine (top support)

- ❏ Eight ⁵⁄₁₆" x 3½" carriage (or machine) bolts, nuts, and washers

- ❏ Two ⁵⁄₁₆" x 4" lag bolts

- ❏ Handful of galvanized 16d nails

- ❏ Handful of galvanized 6d nails

Grain Trough Feeder

Note: You may need an assistant to hold the boards while you nail them.

1. **Bottom.** From the 6-foot piece of fir, cut two pieces 12¾ inches long for the trough ends. Trim the two 8-foot pieces to *exactly* 96 inches. (Lumber sometimes comes a little longer than the specified length. You must trim it to obtain exact measurements.)

2. Nail the sides onto the end pieces, and square up the frame. *Tip:* Drill pilot holes for the nails before you hammer the tray together. Galvanized nails often split the wood when you drive them in. The pilot holes keep this from happening.

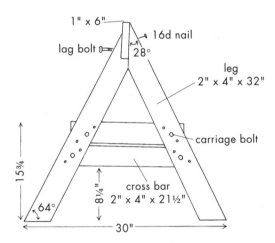

3. Use 6d nails to fasten the plywood (best face down) to the frame.

4. **Legs.** From the 12-foot piece of fir, cut four pieces 32 inches long for the legs. As the diagram shows, cut across the bottom end of each leg at 64 degrees. Cut across the top ends at 28 degrees.

5. *Leg braces.* From the remainder of the 6-foot fir used in Step 1, cut two pieces 21½ inches long. Cut across each end of both pieces at 64 degrees.

6. Align the two legs on a flat surface as they will look when the feeder is fully assembled. There should be 30 inches between the outside tips of the legs at the bottom. At approximately 8¼ inches from the bottom of the legs, lay the leg braces across the legs. Check the angles for match and fit, and lightly tack in place.

7. Drill a ⁵⁄₁₆-inch hole through the bar and the leg, as shown.

8. Disassemble each leg. Cut a notch in the top of one leg as shown, ¾ inch deep to accept the 1" x 6" top brace.

9. Reassemble each set of legs, bolting the bar securely to each leg with the carriage bolts.

10. Lay the feeder tray on the brace of one set of legs. Align and drill through each leg, into and through the end block of the tray. Bolt in place and repeat on other end.

11. Take the piece of #2 pine, and slip it into the notch at the top of each leg set. Once it is in place, use a ⅞-inch spade bit to make a recess for the head of the lag bolt. Use a bit roughly one-third smaller in diameter than the lag bolt to drill a pilot hole for it; do not make the pilot hole as long as the lag bolt. Insert the lag bolt and screw up tightly.

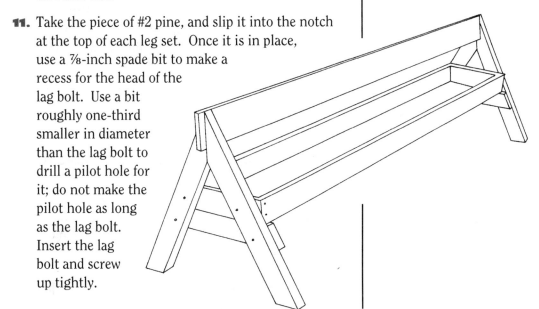

12. Finish fastening the leg on the other side with a 16d nail. Put two more 16d nails at each joint of the leg sets.

13. Drill a ⅜-inch drain hole through the bottom of the tray at each corner to allow rainwater to escape. A common criticism of wooden grain troughs is that they hold water.

Wool Craft Projects

Washing Fleece

Fleece is washed by soaking it in very hot water and plenty of soap or detergent — *without* rubbing or moving it about in the water. If it will be used for handspinning, choose only the best and cleanest parts. For felting, any quality of wool will suffice. Detergent cuts grease better than soap, and it also rinses out more easily. If you are washing a large quantity of fleece (20 pounds, for instance), you will need about 10 cups of detergent for a 20-gallon laundry tub filled to within 5 inches of the top with hot water (the hottest tap water you can get). Do not use your bathtub for this process, because the water will cool down too fast and the grease that the hot water lifts from the fibers will be redeposited on them.

Without taking a lot of time and effort, you want to

- Wash the maximum quantity of wool
- Get the wool very clean of grease and gumminess

Washing the wool. Pull the *skirted* fleece apart, and shake out as much of the seeds and dirt as possible. This also keeps the shorn ends from matting during the washing process. Push as much wool into the hot water as you can get into the tub. Cover the tub to keep the heat in. Soak the wool for 1 to 2 hours.

Squeeze the washed wool as you take it out of the detergent water, and fill two mesh bags or old pillowcases with as much fleece as will fit into your washing machine. Place the bags into your washing machine, and run them through the spin cycle. Repeat with the rest of the washed wool.

If you are going to be washing a great deal of fleece, it would be good to have an old washing machine just for this purpose, rather than using the family machine.

Rinsing the wool. The wool is now ready for rinsing. If you expose the fleece to extreme changes in temperature, it will felt, so it is very important to have the rinse water at the same temperature as the slightly cooled-down wash water, but still hot enough to remove the dissolved gumminess. Discard the wash water from the tub and replace with clean, hot rinse water. This time, add just half of the washed wool to the rinse water. Squish the wool up and down several times, remove it, and put it in the mesh bags to spin out the rinse water in the washing machine. Repeat this procedure with the other half. If the wool is extremely dirty or greasy, you may have to give it two rinses.

Drying the wool. Fluff the wool, and put it out on wire racks to dry. Rust-free chicken wire makes a good drying rack. When it is dry, put it in a cloth bag or a cardboard container, seal well, and store until use. Old pillowcases without holes allow ventillation but keep out moths.

Wool Dyeing

Wool for felting or spinning can be dyed easily at home, using grocery-store ingredients. Wash the wool before you begin. Wet the wool before placing it in whatever dye bath you use.

Kool-Aid Dyeing. In a large jar, dissolve sugarless Kool-Aid in ½ gallon of water. If you use "lighter-colored" flavors, you will need more Kool-Aid; the

Washing Small Quantities of Fleece

Follow the directions for washing large quantities of fleece. Be sure to remove the fleece from the soaking and rinse waters before they cool down. Be sure to handle it gently and don't expose it to sudden changes in water temperature, or it will start to felt.

"vivid-colored" flavors require less. We suggest 4 packets of Lemonade, Pink Lemonade, or Sunshine Punch; 3 packets of Cherry and Lemon-Lime; and 2½ packets of Raspberry, Strawberry, Orange, Tropical Punch, or Black Cherry. Various combinations of flavors can be used also.

Ask for some adult help for this step. Large pans of boiling water can be difficult and dangerous to handle. In a large enamel or stainless-steel pan, heat 1 quart of water. Add the ½ gallon of dye matter and ½ cup of white vinegar. Put in some wet fleece — be sure not to add too much or there won't be room for the water to simmer. Heat the water until it barely simmers, and keep it at that temperature for ½ hour. Remove the pan from the heat and let it cool to lukewarm. Remove the wool from the dye. Rinse it several times in warm water, until the water shows no trace of color. Squeeze out the final rinse water by hand or with the spin cycle of your washer (with permission). Put it on a wire rack to dry. (Chicken wire makes a very good drying rack.)

Crepe-paper dyeing. Bright-colored crepe paper (in rolls, not streamers) also has enough color to dye wool. Simmer the crepe paper in water to extract the dye, add vinegar, and then simmer the wool in the same way you would dye with Kool-Aid.

Commercial fabric dyes. Rit and Tintex and other packaged dyes probably cost a little less than Kool-Aid, and they give more predictable colors. Craft shops sell other brands of fabric dyes in a great range of colors. Follow the instructions on the package.

Introduction to Spinning: Spin without Tools

To get an understanding of the basic spinning process, you can spin wool using nothing but your hands. In fact, one of the earliest forms of spinning was done without any tools, by just rolling wool on the lap with one hand, and drawing it out into yarn with the other hand.

Start with a small handful of wool — either teased (fluffed up) grease wool or gently washed wool. Hold the fibers lightly in one hand. With the other hand, slowly draw a wisp out of the fibers, twisting them as you draw them out. Continue to draw out more fibers, twisting as you do, until you have a good length of actual yarn, still attached to the handful of fibers. This is enough to give you an idea of what spinning is all about; you will even get the feel of the twisting fibers. Each fiber of wool has little scales along it, which you can see only with a microscope; these help make the fibers cling together as it is twisted. Wind a little bit of this spun yarn into a ball, still attached to the puff of wool. Hold the ball of yarn in your hand and repeat the drawing and twisting.

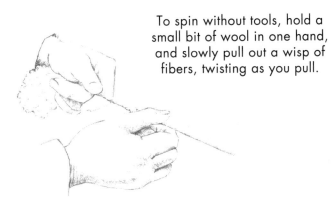

To spin without tools, hold a small bit of wool in one hand, and slowly pull out a wisp of fibers, twisting as you pull.

Spinning with a Spindle

Spinning with a spindle involves the same process as spinning without tools, except that you twirl the spindle and it transfers its twist to the fibers as you draw them out. In other words, the spindle does the wool twisting action, leaving you with both hands free to control the size and texture of your yarn.

Drop Spindle

Use a ⅜-inch dowel or a smooth stick, about 10 to 12 inches long. Whittle a notch close to one end of the stick and a point on the other end. The *whorl* can be a disk of wood about 3 inches in diameter and ½ to 1 inch thick, or a small potato cut in half. Make a hole in the disk or potato into which you can insert the spindle, with about 2 inches of the pointed end sticking through on the other side. If you use a potato, face the sliced side up toward the notched end of the stick.

Twirl the spindle and draw out the fibers as they twist.

In some parts of the world, spinning is still done with a small, rough stick, hooked at the tip. The shorter the fibers, the thinner the stick must be. The stick is twirled to draw wool out of a teased fiber supply that is weighted down with a stone.

To make spindle spinning easier to learn, practice first with commercial yarn before you use your own fibers. Practice how to twirl the spindle to get the feel of its weight and balance.

Beginning to spin. Tie about a yard of *leader* yarn to the spindle, just above the whorl. Wind it around the spindle a few times, then pass it over the whorl, catch it over the bottom of the spindle, bring up again, and make a half-hitch around the spindle tip at the notch. Lay your teased wool fibers against the leader yarn, and twist the spindle in a clockwise direction until the twist of the leader grabs onto the wool. With your right hand (left hand, if you are left-handed), twirl the spindle again (always clockwise), let go of the spindle, and help the other hand to draw out and control the twist. The spindle will continue to rotate gently, inserting more twist as it turns. Do not pull more fibers from the supply than can be held together by the twist. Whenever the spindle slows, reach down and twirl it again. If it loses momentum, it will reverse direction and start to untwist the yarn. Your yarn will then weaken and break.

Winding on. When the yarn is so long that the spindle touches the floor, it is time to wind up the yarn. Loosen the half-hitch and wind the yarn around the spindle, close to the whorl. Wind the yarn into a cone shape, with the widest end against the whorl. By occasionally zigzagging back and forth, you can prevent the newly spun yarn from imbedding itself into yarn already wound on. You can circle the spindle with the hand carrying the yarn, or you may find it more convenient to place the base of the spindle against you, and give it a whirl, winding the yarn around it as it turns.

Wind yarn on the spindle in a cone shape.

Knitting from Fleece

Even if you haven't learned to spin, you can still knit from your own fleeces. While the preparation of the wool may seem tedious, it probably takes about the same amount of time that it would take to card and spin yarn.

The wool for fleece knitting must have a staple length of at least 4 inches. Make bundles of fleece, about 4 inches in diameter. Tie each bundle together firmly (but not too tight) with a strip cut from old nylon pantyhose. Soak the bundles in hot water and detergent for about 1 hour. If you use quite a bit of hot water and fleece and a fairly large pan for the soaking, the water will not cool down too much. Place the bundles in two mesh bags or pillowcases, put them in your washing machines, and use the spin cycle to spin the water out of the bundles. Rinse the bundles in warm water, and again spin out the water. Remove the nylon ties carefully as you place the bundles on a rack to dry.

When the wool is dry, gently pull the wool fibers out of the bundles into a continuous strand. Try to keep the pulled strand a consistent size as you pull. The wool bundles will stay in fairly undisturbed condition if you have handled them carefully. Knit with this strand as you would with a length of spun yarn.

Wool Felting

The first fabrics made by humans were not spun and woven, but felted. Wool (and hair) can be felted because of the overlapping scales on each fiber. When fleece is subjected to a combination of moisture, heat, lubrication, pressure, and movement, the fibers become matted, forming a fabric up to 1 inch thick. This fabric can easily be made into small items such as mittens, caps, and slippers. There is (in

For centuries, shepherds in many countries have worn either felt cloaks or sheepskin tunics.

Felting Equipment

- ❏ Clean fleece
- ❏ Sheet of heavy plastic (such as a shower curtain)
- ❏ Cold water *and* hot (even boiling) water
- ❏ Liquid soap (such as dish detergent)
- ❏ Washing soda (optional)
- ❏ An old-fashioned washboard or handmade felting board (optional)

Bronze Age ruins have contained felt floor coverings that were used 5,000 years ago.

theory) no limit to the size of what you can make, if you work outdoors and have plenty of helping hands. You could even make a piece as large as a football field! In Asia, some nomadic tribes still use tents of felt, which they call *yurts*.

You can use washed wool or grease wool, either teased by hand or carded with hand cards, a drum carder, or a carding slope. Short, fine, crimpy wools felt better than long, coarse, lustrous wools, but any wool can be used to make felt.

Teasing the fibers. Wool clumps must first be either carded or *teased* apart by hand. Teasing is also called *picking,* because you actually pick the tufts of fiber apart until they form a soft, fluffy cloud.

Layering the fleece. The easiest method of feltmaking is to place a sheet of heavy plastic on a table that will not be damaged by hot and cold water. (Or, you can work outdoors on a hard-surfaced walk or driveway.) Take a thin layer of teased wool and place it on the plastic. It should cover a bigger area than you need for the finished project, because the felting process will shrink the piece by as much as one third. Cover the first layer with a second layer, this time with the wool fibers lying in the opposite direction. Keep adding layers, alternating the direction of the fibers with each layer, until the stack is about 1 inch thick when you press down on it. Try to keep the thickness as even as possible across the entire piece so that you don't get lumps and thin places in the completed felt.

Rubbing. Fill a pint jar with *very hot* water, and add about 1 tablespoon of liquid soap. A tablespoon of washing soda will help the process. Pour a small amount of this liquid onto your layers of wool, and press gently with your hands to remove the excess water. It shouldn't be too sloppy wet. Now, start moving your hands in a circular motion over the surface of the wool, pressing gently at first. After a few minutes of this process, carefully turn the wool piece over and rub the other side. Add more soapy water as needed. The soap not only helps the fibers to felt together, but is a lubricant that makes it easier to rub the surface. Some people alternate hot water with cold. Others wait until the wool piece is felted enough to hang together well, then they dunk it alternately in hot water and cold water several times. The piece is felted when you can pinch the surface and not be able to pull fibers from it.

Ornamenting the felt. Once your piece has started felting nicely, you can lay out a design on the surface with pieces of different colored fleece and even dyed wool yarn. Continue to rub the piece, and the pattern will become felted into the finished piece.

Hardening the felt. There are several ways to harden and shrink your felt piece, making it firmer and less prone to future shrinkage. You can roll the felt piece tightly and wind it up with a string until the next day, or you can roll it up and pound it with your fist, a stick, or a hammer. Walking and stomping on the rolled felt is an ancient method of hardening it. Rubbing it on an old washboard or felting board (for directions, see page 102) is a good way to harden small items.

Some Felted Projects

Boot liners. Make flat felt pieces large enough that they can be trimmed into boot insoles and put them in your rubber boots for cold weather.

Slipper soles. Make insoles like the boot liners. Overcast the edges by hand or with a zig-zag sewing machine stitch. Crochet or knit a slipper top onto the sole.

Caps. Many styles of caps and hats can be made by felting. A round shape is best. When it starts to felt, begin the actual shaping. Make it a little thicker in the

Arrange thin layers of wool in a thick stack.

Pour hot, soapy water over the layers and rub in a circular motion.

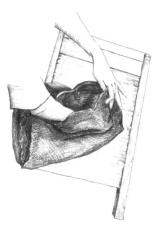

Harden felt by rubbing it on a washboard.

middle than at the edge, and then put the piece over a kitchen bowl to form the cap shape. Work the center portion gently onto the bowl, rubbing to stretch and shape it. At the same time, keep working the edge to felt (and shrink) it. When you are done, trim the edge with scissors.

Mittens. When you create your first pair of mittens, make two pieces of felt for each mitten — a top and a bottom. On a piece of paper, trace around your fingers and thumb. Cut ½"–¾" outside the tracing line and use as a pattern to cut four pieces of felt. Once you have some experience in mitten making, try to make each mitten in one piece. There are excellent directions for a nifty method of making these in the Spring 1988 issue of *SpinOff* magazine. (Back issues of *SpinOff* can be ordered from sources listed in the Appendix.)

Larger items. Flat pieces can be made into rugs and saddle blankets.

FELTING BOARD

Glue the molding pieces onto the plywood, spacing them evenly over the entire board, to create a washboard-type surface. Instead of the wooden molding, you can use a ribbed piece of rubber floor runner.

Materials

- ❏ 18" x 18" piece of plywood

- ❏ Twenty-four 18-inch long pieces of ½", ⅝", or ¾" half-round molding

- ❏ Waterproof glue or contact cement

Sheep Calendar

(Adapted from The Sheep Calendar created by and available from the Norse-man Sheep Company; see page 109 for address.)

This checklist is meant to help you remember what needs to be done to care for your sheep throughout the year. The page numbers tell where in the book you can find more information about each task. The calendar starts with May, as this is the month most new shepherds begin their flocks.

May

- ❏ Check your fences *before* bringing any new sheep home (pages 23-24).
- ❏ Check your pasture for toxic plants (pages 30-31).
- ❏ Give new sheep hay *before* turning them onto more lush pasture than they are used to (pages 33-34).
- ❏ Check for ticks, and treat them if you see even *one* (pages 44-45).

June

- ❏ Practice pasture rotation (pages 23-24).
- ❏ Provide plenty of water, salt, and vitamin/mineral mix (pages 29-30).
- ❏ Watch for fly strikes (clusters of tiny eggs) and maggots.
- ❏ Keep rear ends trimmed to discourage flies.

July

❑ Provide shade for your sheep in very hot weather.
❑ Provide cool, fresh water — 1–2 gallons a day for adult sheep.
❑ Check the feet of limping sheep, and trim hooves, if necessary (pages 46-48).
❑ Worm your sheep and record date of worming (pages 43-44).

If you are breeding your sheep:

❑ Put your ram in a shaded pasture next to the ewe.
❑ Shear the ram's scrotum to keep him cool.
❑ Start flushing the ewe before you plan to turn the ram in for breeding (pages 61-62).
❑ Do not let your ewe eat clover; it decreases her fertility.

August

❑ Continue to provide shade.
❑ Worm now, if you didn't do it in July (pages 43-44).
❑ Buy grain for winter and store in rodent-proof containers, where sheep cannot possibly break into it.

If you are breeding your sheep:

❑ For early lambs, keep the ram in a shady place during the day, and turn him in with the ewe in the evening.
❑ Do not let your ewe eat clover; it decreases her fertility.
❑ Give the ram a small amount (¼–½ pound) of grain daily.
❑ Keep the ram with your ewe at least 6 weeks so that there are at least two chances for mating (pages 63-64).
❑ Mark on your calendar the date you turned the ram in with the ewe, so that you will know the earliest date you could expect the lambing (page 63). Also mark the date if you observe any mating take place.

September

- ❑ Locate your winter supply of hay.
- ❑ Keep water, salt, and mineral/vitamin supplement always available.
- ❑ Make a list of repairs needed on shelter, fencing, and equipment, and start your repairs before cold weather.
- ❑ If you are worming, be sure to read the label regarding any precautions about timing prior to slaughter.
- ❑ Get sizeable lambs to market. Shear them first, or ask for their pelts.
- ❑ Clean out the place where you plan to store winter hay; save the manure you collect and spread it on the vegetable garden.
- ❑ If you have apples, feed a few windfalls (but not too many at one time) to sheep. Set some windfalls aside for winter.

If you are breeding your sheep:

- ❑ For late lambs, flush ewe and then turn in the ram (pages 61-64).
- ❑ Reduce grain gradually after flushing (pages 61-62).
- ❑ Record dates when you see breeding take place.

October

- ❑ Clean out sheep sheds and/or barn, and spread sheep manure on garden.
- ❑ Get in the winter hay.
- ❑ Clean out and check your waterers; winterproof your faucets.

If you are breeding your sheep:

- ❑ If you have seen no signs of breeding, consider borrowing a ram. Breeding time may be running short now for most breeds.
- ❑ Check over your lambing supplies (pages 64-66). Mail order supplies, if necessary (see pages 109-110 for addresses).
- ❑ Make lamb jugs (pages 87-90).

November

- ❏ Keep grain in rodent-proof containers, and take steps to get rid of rodents.
- ❏ Order antibiotics and store them in a refrigerator for emergencies.
- ❏ If your sheep is limping, check hooves and trim, if necessary (pages 46-48).

If you have bred your sheep:

- ❏ Check your lambing supplies (pages 64-66). Order ear tags, if you plan to use them (pages 70-71; see page 109 for sources).
- ❏ Put the ram in a separate area from the pregnant ewe, so that he doesn't injure her.
- ❏ If the ram is run down, be sure to feed him well.
- ❏ Add a small amount of stock molasses to the pregnant ewe's drinking water.
- ❏ Make lamb jugs if you haven't already (pages 87-90).

December

- ❏ Put molasses in ewe's drinking water — it helps keep the water from freezing.

If you have bred your sheep:

- ❏ Crotch ewe to prepare her for lambing (page 62). Also, remove dirty tags from udder and legs.
- ❏ Begin checking the udder of a very pregnant ewe; if it is hard and lumpy, she may have no milk and you should be prepared to bottle-feed her lamb (pages 34-35).
- ❏ Four weeks before lambing time, begin feeding grain (¼–½ pound daily) and some hay to ewe (page 62).
- ❏ If your ewe is listless, she may have pregnancy toxemia; call your veterinarian (page 66).
- ❏ Give calcium supplement to your pregnant ewe.

January

If you have bred your sheep:

- ❑ Watch pregnant ewe carefully for signs of labor (pages 66-68).
- ❑ Be sure pregnant ewe is getting exercise (page 62).
- ❑ If ewe refuses to eat, it may be a sign of pregnancy toxemia or just a sign that she is about to lamb.
- ❑ Crotch ewe if you haven't yet (page 62).
- ❑ Add molasses to pregnant ewe's drinking water.
- ❑ After lambing, put the ewe and lamb in the jug (pages 64-65). Strip the ewe's teats, and "snip and dip" the lamb's umbilical cord (page 69). Give ewe warm molasses water.
- ❑ Dock lambs' tails when they are 2–3 days old, and castrate male lambs when they are about 10 days old (pages 39-41).

February

- ❑ Make sure salt is available.

If you have lambs:

- ❑ Watch baby lambs to be sure that they are having normal bowel movements.
- ❑ Watch twin lambs to be sure one isn't growing more rapidly than the other; supplement feed for slower-growing lamb.
- ❑ If a ewe loses a lamb, encourage her to adopt an orphan lamb (pages 75-77).
- ❑ Dock lambs' tails when they are 2–3 days old, and castrate male lambs when they are about 10 days old (pages 39-41).
- ❑ Ear tag lambs (pages 70-71).
- ❑ Check ewe's feet, and trim and treat them, if necessary, before turning her out of the pen (pages 46-48).
- ❑ Give plenty of fresh, clean water.
- ❑ Continue feeding grain to nursing ewes (page 70).
- ❑ Prepare the lamb creep (page 90).

March

- ❑ Worm all sheep. Restock worming supplies (pages 43-44).
- ❑ Start shearing, if weather permits (pages 49-54). Keep mothers with their lambs as much as possible, to avoid confusion. Do not shear wet sheep. Keep fleeces clean.
- ❑ At shearing time, or 10 days later, treat for ticks (pages 44-45).
- ❑ Trim hooves at shearing time (pages 46-48).
- ❑ Clean out the barn or shed, and put old hay and manure on the vegetable garden or on an area of the pasture that needs to be fertilized.

April

- ❑ Locate your feeder on well-drained ground to avoid hoof trouble.
- ❑ If you didn't do it in March, clean out the barn or shed, and put old hay and manure on the vegetable garden or on an area of the pasture that needs to be fertilized.

If you have lambs:

- ❑ Keep fresh water, salt, and feed in the lamb creep (page 90).
- ❑ Before turning ewe onto new pasture, let lambs in first, so that they get the best of the grass.
- ❑ Worm lambs when they weigh about 40 pounds (pages 43-44). You may want to wean them at this time (pages 37-38).
- ❑ Before lambs are weaned, decrease ewe's grain ration and feed her on hay alone (page 78).

Helpful Sources

Sheep Supplies

Sheepman Supply Co.
P.O. Box 100
Barboursville, VA 22923
Catalog 50
Large variety of equipment and supplies for sheep

C.H. Dana Co.
Hyde Park, VT 05655

Nasco Farm and Ranch
901 Janesville Ave.
Ft. Atkinson, WI 53538
or
Nasco West
1524 Princeton Ave.
Modesto, CA 95352

Norseman Sheep Co.
Route 1, Box 141A
Wellsville, KS 66092
*Computer program for sheep; newborn lamb carrier
Send long SASE for information*

Peterson's Natures Way
Route 2-2584
Sidney, MT 49228
Lamb adoption coat, colostrum powder, other supplies

Protein Technology
701 4th Ave. South, Suite 1350
Minneapolis, MN 55415
Colostryx milk whey antibody product for lambs

Mid-States Livestock Supplies
3900 Groves Rd., Box 32461
Columbus, OH 43232-0461

Good Shepard Products
4052 State Hwy. 38
Drain, OR 97435

Premier Sheep Supplies
Box 89
Washington, IA 52353

D-S Livestock Equipment St.
Star Route Box 20
Frostburg, MD 21432

Sheep Publications

sheep! Magazine
W 2997 Markert Road
Helenville, WI 53137

The Sheep Producer
Route 2, Box 131A
Arlington, KY 42021

The Southeastern Sheepman
P.O. Box 350
Loganville, GA 30249

The Shepherd
5696 Johnston Road
New Washington, OH 44854

Sheep Canada
Box 113
Moosehor, MB R0C 2E0
Canada

The Black Sheep Newsletter
Route 1, Box 288
Scappoose, OR 97056

Dorset Digest
Arnold Farm Publications
Route 1, Box 169
Sheldon, WI 54766

Sheep Videotapes

Rural Route Videos
Box 113
River John, Pictou Co., NS
B0K 1N0 Canada

"Raising Bonue Lambs"
N. Smith, National Marketing
Land O' Lakes Inc.
2827 8th Avenue South
Fort Dodge, IA 50501

Lincoln University Cooperative
Extension, Media Dept.
900 Moreau Drive
Jefferson City, MO 65101

Lamb Cookbooks

The Black Sheep Newsletter Lamb Cookbook
Route 1, Box 288
Scappoose, OR 97056

Great Sausage Recipes and Meat Curing by Kutas
(available from Norseman Sausage Supplies, Route 2, Box 141A, Wellsville, KS 66092; *free pamphlets on lamb cookery*

The American Sheep Industry Assoc.
6911 S. Yosemite Street
Englewood, CO 80112-1414

Books on Sheep Raising

Most sheep books may be mail ordered from *sheep! Magazine,* W 2997 Markert Road, Helenville, WI 53137. Roger Pond's books are available from Pine Forest Publishing, 314 Pine Forest Rd., Goldendale, WA 98620; send SASE for publications list.

Beginning Shepherd's Manual
by Barbara Smith (Iowa State University Press, 1983)

The Black Sheep Newsletter Companion (articles from the first five years of The Black Sheep Newsletter, Route 1, Box 288, Scappoose, OR 97056)

Health Care Practices (available from William Kruesi, R.D. 1, Box 549, Wallingford, VT 05773; send SASE for publication list)

It's Hard to Look Cool When Your Car's Full of Sheep by Roger Pond (Pine Forest Publishing, 1989)

Livestock Showman's Handbook by Roger Pond (Pine Forest Publishing, 1986)

Raising Sheep the Modern Way by Paula Simmons (Garden Way Publishing, revised 1989)

The Sheep Book by Ron Parker (Ballantine, 1984)

The Sheep Raiser's Manual by William Kruesi (Williamson Publishing, 1985)

Tanning

"Home Tanning Methods" Leaflet #21005 (free from Co-op Extension Service, University of California, Davis, CA 95616)

Tan Your Hide by Phyllis Hobson (Garden Way Publishing, 1977)

Information On Electric Fences

Parker-McCrory Manufacturing Co.
2000 Forest Avenue
Kansas City, MO 64108

Premier
Box 89
Washington, IA 52353

Building Plans

"Plans for Farm Building and Equipment" (available from Extension Service, Utah State University, Logan, UT 84321)

"Sheep Handbook of Housing and Equipment" (available from Midwest Plan Service, Iowa State University, 122 Davidson Hall, Ames, IA 50011)

Plans for building a "Lamb Box," docking, giving shots, etc. (available from Hillside Farms, 8351 Big Lake Road, Clarkston, MI 48016)

Building Small Barns, Sheds, and Shelters by Monte Burch (Garden Way Publishing, 1983)

How to Build Small Barns & Outbuildings by Monte Burch (Garden Way Publishing, 1992)

Shearing

Self-Teaching Shearing Chart (available free from *sheep! Magazine,* W 2997 Markert Road, Helenville, WI 53137)

Sheep Coats

Ewe's Cottage
P.O. Box 183
Roundup, MT 59072

Coal Creek Sheep and Wool
7542 Coal Creek Road
Superior, CO 80027

Powell Sheep Co.
P.O. Box 183
Ramona, CA 92065

Northfield Meadows
Route 1, Box 110
Dalbo, MN 55017

Spinning and Weaving Publications

A fine selection of craft books is available from **The Unicorn,** 1338 Ross St., Petaluma, CA 94952-1191 (catalog, $1)

SpinOff
Interweave Press
201 East 4th Street
Loveland, CO 80537

Handwoven
Interweave Press
201 East 4th Street
Loveland, CO 80537

Shuttle, Spindle and Dyepot
120 Mountain Avenue, B101
Bloomfield, CT 06002

The Heddle
P.O. Box 1906
Bracebridge, ON P0B 1C0
Canada

Australian Locker Hooking
by Signe Nickerson (available from The Crafty Ewe, Box 33, Stevensville, MT 59870)

"Felting" (a pamphlet available from Louet, R.R. 4, Prescott, ON K0E 1T0, Canada)

Cooperative Extension Service

For information about sheep and youth programs in your area, write to the Cooperative Extension Service contact listed for your state.

Alabama
D.M. Gimenez, Jr.
Auburn University, AL 36849

Alaska
Ken Krigg
University of Alaska
Fairbanks, AK 99775-5022

Arizona
Ed LeViness
2400 S. Milton Rd.
Flagstaff, AZ 86001

Arkansas
George V. Davis Jr.
University of Arkansas
P.O. Box 391
Little Rock, AR 72203

California
Kim Ellis
University of California
Davis, CA 95616

Colorado
Steve LeValley
Colorado State University
Fort Collins, CO 80523

Connecticut
Louis A. Malkus
University of Connecticut
Storrs, CT 06268

Delaware
Richard Barczewskie

University of Delaware
  Substation
RD. 1, Box 658
Dover, DE 19901

Florida
David L. Prichard
University of Florida
231 Animal Science Blvd.
Gainesville, FL 32611

Georgia
Calvin Alford
Animal Science Cooperative
  Extension Service
University of Georgia
  Colliseum
Athens, GA 30602

Guam
Manual V. Duguies
College of Agriculture and
  Life Sciences
University of Guam, Mangilo
96923

Hawaii
Brent A. Buckley
University of Hawaii
Honolulu, HI 96822

Idaho
Gerald T. Schelling
Dept. of Animal Science
University of Idaho
Moscow, ID 83843

Illinois
Gary S. Rirketts
University of Illinois
328 Mumford Hall
1301 W. Gregory Drive
Urbana, IL 61801

Indiana
J.C. Forrest
Purdue University
West Lafayette, IN 47907

Iowa
Dan Morrical
Iowa State University
Kildee Hall
Ames, IA 50011

Kansas
Clifford Spaeth

Kansas State University
Manhattan, KS 66506

Kentucky
G.L.M. Chappell
University of Kentucky
Agricultural Science Center
  South
Lexington, KY 40546

Louisiana
Terry L. Deemas
Louisiana State University
226 Knapp Hall
Baton Rouge. LA 70803

Maine
Robert Hugh
University of Maine
332 Hitchner Hall
Orono, ME 04469

Maryland
Scott M. Barao
University of Maryland
College Park, MD 20742

Massachusetts
Joseph Tritschler
University of Massachusetts
301 Stockbridge Hall
Amherst, MA 01003

Michigan
Margaret Benson
Michigan State University
Anthony Hall
East Lansing, MI 48824

Minnesota
Robert M. Jordan
University of Minnesota
121 Peters Hall
St. Paul, MN 55108

Mississippi
Joe P. Baker
Mississippi State University
Box 5425
Mississippi State, MS 39762

Missouri
Helen Swartz
University of Missouri,
  Lincoln
900 Moreau Drive
Jefferson City, MO 65101

Montana
Rodney Kott
Montana State University
Bozeman, MT 59717

Nebraska
Ted H. Doane
University of Nebraska
Lincoln, NE 68583

Nevada
A.Z. Joy
University of Nevada
Box 210, Court House
Ely, NV 89301

New Hampshire
F. Carlton Ernst
University of New Hampshire
209 Kendall Hall
Durham, NH 03824

New Jersey
Ovine Research Extension
  Service
Rutgers State University
New Brunswick, NJ 08903

New Mexico
James M. Sachse
New Mexico State University
Drawer 3AE
Las Cruces, NM 88003

New York
D.E. Hogue
Cornell University
Frank B. Morrison Hall
Ithaca, NY 14853

North Carolina
Roger G. Chickenbarger
North Carolina State
  University
Box 7621
Raleigh, NC 27695-7621

North Dakota
Roger Haugen
North Dakota State University
Fargo, ND 58105

Ohio
Steve Baertsche
Ohio State University
2029 Fyffe Rd.

Columbus, OH 43210-1095

Oklahoma
Greg A. Highfill
Oklahoma State University
316 E. Oxford
Enid, OK 73701

Oregon
Tom Bedell
Oregon State University
204 Withycombe Hall
Corvallis, OR 97331

Pennsylvania
Claire C. Eagle
324 William L. Henning Bldg.
University Park, PA 16802

Puerto Rico
David Jimenez
University of Puerto Rico
Call Box 21120
Rio Piedras, PR 00928

Rhode Island
Walter A. Gross
University of Rhode Island
Kingston, RI 02881

South Carolina
R.F. Clements
Clemson University
P.O. Box 640
Abbeville, SC 29620

South Dakota
James M. Thompson
South Dakota State University
Animal and Range Science
  Dept.
Box 2170
Brookings, SD 57007

Tennessee
James B. Neel
University of Tennessee
P.O. Box 1071
Knoxville, TN 37901

Texas
Texas A & M University
College Station, TX 77843
and George A. Ahlschwede
Route 2, Box 950
San Angelo, TX 76901

and Frank Craddock
Drawer 1849
Uvalde, TX 78801

Utah
Clell Bagley
Utah State University
Logan, UT 84322-4815

Vermont
Ed McGarry
Vermont Agricultural Center
North Hero, VT 05474

Virgin Islands
Paul Boateng
College of Virgin Islands
P.O. Box L
Kingshill, St. Croix 00850

Virginia
Steven H. Umberger
Virginia Polytechnical
  Institute and State Univ.
Blacksburg, VA 24061

Washington
Ladd A. Mitchell
Washington State University,
  Court House
Ephrata, WA 98823 or
Washington State University
Pullman, WA 99164

West Virginia
Richard M. Koes
West Virginia University
Morgantown, WV 26506

Wisconsin
Sherwood Lane
University of Wisconsin
1675 Observatory Drive
Madison, WI 53706

Wyoming
Hans Nels
University of Wyoming
Box 3354, University Station
Laramie, WY 82071

District of Columbia
Gary M. Weber
USDA Extension Service
Room 3334, South Building
Washington, D.C. 20250-0900

Glossary

abomasum (n.). A sheep's stomach, where food is digested after it has passed through the three "fore-stomachs."

anemia (n.). In sheep, a blood loss caused by blood-feeding worms.

antibiotic (n.). A type of medicine that destroys organisms that cause disease.

antibodies (n.). Substances in the blood that naturally help fight disease.

bellwether (n). A castrated male sheep that wears a bell and is the leader of a flock.

block (v.). Comb and trim a sheep's coat after shearing to make it smooth.

bolus (n.). A large pill.

bottle jaw (n.). Swelling under the jaw of sheep caused by a heavy infestation of worms.

breed (n.). A type of animal that has specific inherited characteristics.

breed (v.). Mate and produce offspring.

bummer lamb (n.). An orphan lamb that is fed with a bottle rather than nursing from a ewe.

calcium deficiency (n.). Illness caused when a ewe does not have enough calcium for her own body and for her lambs. Calcium is a mineral that helps make strong bones and teeth in lambs and other animals.

carcass (n.). The body of a slaughtered animal.

castrate (v.). Remove the testicles of a male animal to permanently prevent it from breeding.

Clostridial diseases (n.). Diseases, such as tetanus and enterotoxemia, caused by bacteria called Clostridium.

club lamb (n.). Lamb raised as a project for 4-H, FFA, or other club.

colic (n.). Severe pain in the abdomen caused by trapped gas.

colostrum (n.). First milk after birth. Formed in the last weeks of pregnancy, it is rich in fat, protein, and protective antibodies.

conformation (n.). Shape and proportions of an animal.

contaminated (adj.). Not clean; containing bacteria, parasites, or other harmful organisms.

contractions (n.). Strong muscle movements in the mother's uterus that push the baby out at birth.

creep (n.). Special area for feeding hay and grain to lambs who are not yet weaned; also, the specially formulated feed given to lambs

crimp (n.). Wavy pattern in individual strands of wool.

cross-fencing (n.). Temporary fences that divide a large pasture into smaller sections.

crossbreed (n. or adj.). An animal whose parents are of different breeds.

crotch (v.). Trim the wool around the dock and udder of a ewe. Also called **Tag** or **crutch.**

cud (n.). Food returned from the first "fore-stomach" to be chewed again.

dental pad (n.). Hard line of gum which sheep have instead of upper front teeth.

diarrhea (n.). Excessive and messy bowel movements or droppings. See also **Scours.**

disinfect (v.). Clean so that germs are killed.

dock (n.). Rear portion of a sheep.

dock (v.). Shorten by cutting, as the tails of sheep.

domestic animal (n.). Animal that has been tamed by humans.

enterotoxemia (n.). Condition caused by overeating or an imbalance in food.

esophagus (n.). Tube that carries food from the mouth to the stomach.

estrous cycle (n.). Reproductive cycle, which in the ewe is normally 17 days long.

estrus (n.). Time during which the female is in heat and can be bred. A ewe is usually in estrus for about 28 hours. If she is not bred in that time, her cycle continues, and she will come into estrus again in 17 days.

ewe (n.). Mature female sheep.

exotic (adj.). Very unusual and striking.

felt (n.). Fabric made of layers of wool pressed and matted very tightly together.

fertility (n.). Ability of the female to produce healthy eggs and become pregnant, and the ability of the male to produce healthy sperm.

fit (v.). Groom a sheep before a show.

fleece (n.). Wool coat of a sheep.

flock (n.). Group of sheep that live together.

flush (v.). Feed ewes a highly nutritious diet before they are bred to aid in the reproductive process.

footrot (n.). Painful and crippling infection that affects the feet of sheep.

forage (n.). Food obtained from grazing in a pasture.

gestation (n.). Time required for the

growth and development of the baby from conception to birth. Also called **pregnancy.** In sheep, it averages 147 to 153 days (about 5 months).

graft (v.). Have a ewe accept and serve as mother to a lamb that is not hers.

grease wool (n.). Unwashed fleece of a sheep.

halter (n.). Rope or leather strap that fits around the head and neck of an animal.

handspin (v.). Twist and draw out fleece to form yarn.

immunization (n.). Process of using medicines to prevent sheep from getting certain diseases.

infestation (n.). Presence of a large number of harmful parasites or insects.

instinct (n.). Animal's natural tendency to behave in a certain way.

intranasal (adj.). Administered through the nose, as medicine.

jug (n.). Pen for ewes and their newborn lambs.

ked (n.). Parasite, also called a *sheep tick,* that sucks the blood of sheep.

labor (n.). Process of giving birth, during which contractions push the baby out; for ewes, usually a period of one to three hours.

lambing (n. or v.). Giving birth to lambs.

lice (n.). Parasite that bites or sucks the blood of animals, causing severe itching.

livestock (n.). Domestic animals, such as sheep, cattle, and horses.

maggots (n.). Flies in a stage of their lives when they resemble fat worms.

manure (n.). Animal droppings.

market lamb (n.). Lamb raised to be sold, usually for meat.

mastitis (n.). Infection of the udder.

membrane (n.). Very thin covering.

nits (n.). Eggs and babies of insects, such as flies.

nurse (v.). Suck milk from the udder.

omasum (n.). Third "fore-stomach" in a sheep's digestive system, where excess water is removed from food.

parasite (n.). Harmful organism, such as a tick, that lives on another organism.

pastern (n.). Ankle joint just above the hoof of a sheep.

pelt (n.). Skin of a sheep or other animal with the wool or fur still on it.

pesticide (n.). Chemical used to kill pests, such as insects.

polled (adj.). Having horns.

predator (n.). Animal that hunts other animals for food.

pregnancy toxemia (n.). Life-threatening disease of pregnant ewes.

pregnant (adj.). Having a baby growing inside

purebred (adj.). Bred from many generations of parents of the same breed.

ram (n.). Uncastrated adult male sheep.

registered (adj.). For a purebred sheep, records that have been filed with the appropriate breed society.

reproductive diseases (n.). Diseases that happen during and as the result of pregnancy.

respiratory diseases (n.). Diseases that affect the ability to breathe.

reticulum (n.). Second of the three "fore-stomachs" in a sheep's digestive system.

rotation (n.). Process of alternating use, such as pastures.

rumen (n.). First of the three "fore-stomachs" in a sheep's digestive system.

ruminants (n.). Animals, including sheep, cattle, goats, and deer, whose digestive system makes it possible for them to feed on grass and hay.

scald (n.). Skin infection between sheep's toes.

scours (n.). Yellow diarrhea in lambs usually caused by overeating.

scrotum (n.). In male animals, the sac that contains the testicles.

second cuts (n.). Short lengths of wool created by running the shears over a place on a sheep that has already been sheared.

shear (v.). Remove the heavy wool coat of sheep.

skirt (v.). Remove the undesirable wool from the outside edge of a sheared fleece.

tag (v.). See Crotch.

teats (n.). Fingerlike projections on the udder (bag) through which milk flows to the baby.

temperament (n.). Usual way a person or animal behaves.

testicles (n.). Glands of male animals that create sperm and make reproduction possible.

toxic (adj.). Poisonous.

trait (n.). Inherited characteristic.

twin (v.). Give birth to twins.

udder (n.). Organ that produces milk; also called the **bag.**

umbilical cord (n.). Long cord that connects the mother and the developing baby in the uterus. During gestation, it transports nutrients from the mother to the baby and wastes from the baby to the mother.

uterus (n.). Hollow organ with muscular walls where the fertilized egg attaches, develops, and grows during gestation. Sheep have a two-sided uterus.

vaccine (n.). Substance, usually given in an inoculation, that prevents certain illnesses.

ventilated (adj.). Constructed so that fresh air can circulate easily.

wean (v.). Change a baby's way of feeding from nursing to eating other food; separate the baby from its mother.

wether (n.). Castrated male sheep.

INDEX

Stomach, 33–34
Storing the fleece, 52
Suffolk sheep, 12–13

T
Tagging. *See* Crotching; Ear
    tagging
Tags, 51
Tails, *See* Docking tails
Tallow candles, 79–80
Tanning hides, 110
Teats, 66
    after delivery, 69
Teeth, 18, 52. *See also* Dental
    pad
    health problems, 18
Temperament. *See also*
    specific breed
    of ram, 1, 63
Temperature
    of newborn lambs, 72–73
Ticks. *See* Keds
Tire feeder, 91
Toe glands, 48
Toxemia. *See* Enterotoxemia;
    Pregnancy toxemia
Toxic plants, 30–31, 103

Training, 25–29, 32
    for showing, 55–57
Tunis sheep, 15–16
Twinning, 6, 21
    newborn lamb problems,
    73–74

U
Udder, 19, 66
    mastitis, 72
    problems with, 74, 106
Umbilical cord, 67, 69
Uterus, 67

V
Vaccinations, 41–43
    buying a sheep, 20–21
Videotapes about sheep, 109
Vitamins. *See* Mineral/
    vitamin supplement

W
Wandering lamb, 75
Washing fleece, 94–95
Water, 24, 30
Water bag. *See* Birth sac
Weaning, 37, 78, 108

orphaned lambs, 37–38
Weather, 5, 7
Weaving, 79
    publications, 110
White face, 10–11
Whorl (spindle), 97, 98
Wooden, Uncle Dick,
    Wild West shepherd, 3
Wool. *See also* Fleece craft
    projects, 94–102
    properties of, 49–50
    staple length, 81
Wool cards, 56–57
Wool pool, 80
Worming, 43–44, 104
    buying a sheep, 21
    healthy sheep, 32, 43
    newborn lambs, 104, 108
    and slaughter precautions,
    105
Worms, 44. *See also* Bottle
    jaw; Potbelly
Woven wire fence, 23

Y
Yurts, 100